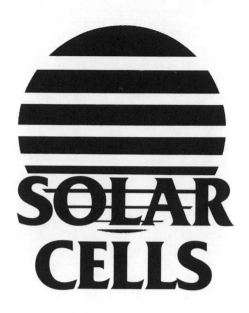

SOLAR CELLS

Their Optics and Metrology

SOLAR CELLS

Their Optics and Metrology

(Optika i metrologiya solnechnykh elementov)

M. M. Koltun

All-Union Scientific Research Institute of Power Sources
(Moscow)

ed. **N. S. Lidorenko**

translated by **S. Chomet**

Allerton Press, Inc. / New York

Library of Congress Cataloging-in-Publication Data

Koltun, Mark Mikhaĭlovich.
  [Optika i metrologiia solnechnykh ėlementov. English]
  Solar cells : their optics and metrology / M.M. Koltun ; ed., N.S.
Lidorenko ; translated by S. Chomet.
    p. cm.
  Translation of : Optika i metrologiia solnechnykh ėlementov.
  Bibliography: p.
  ISBN 0-89864-034-2
  1. Solar cells.   I. Lidorenko, N. S. (Nikolaĭ Stepanovich)
II. Title.
TK2960.K64 1988
621.31'244--dc19   88-4570

Printed in the United States of America

Contents

3 SOLAR CELLS WITH OPTICAL COATINGS   136

4 DETERMINATION OF THE EFFICIENCY AND METRO-LOGICAL PARAMETERS OF SOLAR CELLS AND BATTERIES   195

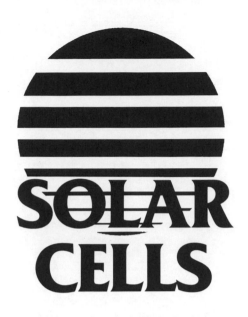

SOLAR
CELLS

Their Optics
and Metrology

Introduction

Solar-energy utilization has now attracted the attention of specialists in different scientific disciplines, ranging from chemistry and biology to solid state physics [1]. Considerable advances have been made in the development of semiconductor solar cells and batteries [2]. Solar batteries for spacecraft are becoming larger, lighter, and cheaper [3], and the range of their terrestrial applications is expanding [4]. At the same time, the efficiency and variety of the batteries are increasing [5].

It is now nearly thirty years since the early foundation-laying work on photoelectric energy conversion appeared [6,7] (its practical and scientific significance is reviewed in [8,8a,8b]). These studies were, in turn, based on the systematic theory of photoelectric phenomena in semiconductors, established in the 1930s and 1940s. In the USSR, this branch of semiconductor physics has developed as a result of the pioneering work of Academician. A. F. Ioffe and his school, who extended our understanding of the nature of photoconductivity and photoelectric phenomena in semiconductors and semiconducting p–n junctions [9–12]. Work done in the Soviet Union and elsewhere on the development of photoelectric devices converting radiant energy, including solar energy, into electric power has been systematically reviewed in specialist monographs [12,13]. Useful theoretical and practical recommendations on the choice of the optimum design of photoconverters and a description of their electrophysical characteristics are given in [14–19]. The development of reliable and low-weight designs for solar batteries (largely for applications in space), and questions relating to the effect of cosmic radiation upon them, are discussed in [20–22], while the development of optimum transmitting, thermally regulating, and protecting optical coatings for solar cells and batteries is discussed in [23].

The present monograph is devoted to the optics and metrology of solar cells, and is the first of its kind. The words "optics" and "metrology" do not appear fortuitously one after another in the title of this book. Detailed knowledge of the optical and metrological characteristics of solar radiation, solar simulators, and semiconductor solar cells, as well as the standardization of numerical values of solar-cell parameters, are the essential prerequisites for estimates of the efficiency of these cells and for further attempts toward the improvement of efficiency. There is no way of estimating the quality of solar-energy converters (including the quality of solar cells) other than measurement by reasonable and reliable methods. The well-known physicist S. Tolanskii has pointed out that the ancient formula "seeing is believing" can no longer be regarded as valid [24]. "Believing" must now follow accurate measurement.

The author is greatly indebted to his editor, Corresponding Member of the USSR Academy of Sciences N. S. Lidorenko, to researchers in various organizations in the USSR, to colleagues and collaborators, and to graduate students (many of the results obtained by collaborative research with them form the basis of much of the work reported on the following pages). G. A. Gukhman, I. V. Gracheva, and T. N. Fedorovskaya participated extensively in the preparation of the manuscript for the press, and to them the author expresses his grateful thanks.

1

Optical and Photoelectric Properties of Solar Cells

The advent of the solar cell was preceded by detailed studies of the optical properties of semiconductors, including the interaction of light with semiconducting materials, which produces nonequilibrium excess charge carriers.

To understand the optical and photoelectric properties of solar cells, we must review, if only briefly, the band structure of semiconductors and how it differs from the electronic structure of metals and insulators. We must also examine the optical properties of semiconducting media, which are determined by the band structure. It is only then that we can proceed to an analysis of the optical and photoelectric properties of solar cells themselves (and, of course, methods of measuring them) as well as the techniques used to determine and investigate the structural and electrophysical parameters of individual semiconducting layers in such cells. These parameters largely determine both the characteristics and the efficiency of solar cells and batteries.

In this chapter, as elsewhere in the book, we shall not be concerned with problems encountered in the concentration of solar radiation or the development of beam splitting devices that modify its spectral composition. These questions have already been adequately reviewed in specialist literature.

1.1 Semiconductors, dielectrics, and metals

When a solid such as a semiconducting crystal is formed, its atoms approach one another to such an extent that their outer electron shells are found to overlap. Instead of the individual orbits of separate atoms we then have to consider collective orbits in which atomic subshells combine into bands that characterize the crystal as a whole. This is accompanied by a radical change in the motion of the electrons: electrons occupying a particular energy level in a given atom acquire the facility of transferring without loss of energy to a similar level in a neighboring atom, and thus freely move throughout the crystal.

The inner shells in isolated atoms and, consequently, those in crystals too, are completely filled. However, the uppermost band, formed from levels occupied by valence electrons, is not always completely filled. The electrical conductivity of crystals, and their optical and many other properties, are largely determined by the extent to which this *valence band* is filled, and by the separation between this band and the uppermost band, which is referred to as the *conduction band.* Electrons arriving in the conduction band from the valence bands (for example, as a result of thermal or optical excitation) can participate in the transport of electric charge, i.e., electric current. The transport of electrons from the point in the valence band at which they are liberated produces the motion of positive charges, called holes, in the opposite direction. Positive charge is always produced in the valence band after the departure of an electron, since until then the band was electrically neutral.

Materials in which the valence band is completely filled, and the separation from the next band is large, are called *dielectrics*. A different energy structure is characteristic for *metals*: the valence band is filled either partially, or it overlaps the next free band, the conduction band. However, when the valence band of a given material is completely filled, but the energy separation from the conduction band is small (conventionally less than 2 eV), the material is referred to as a *semiconductor*. The electrical conductivity and other properties of semiconductors are very dependent on external conditions, especially the temperature T. As the temperature T increases, there is an exponential increase in the number of thermally excited electrons crossing the *band gap* (of width E_g) between the valence band and the conduction band, the number of electrons in the conduction band and of holes in the valence band increases, and the electrical conductivity of the semiconductor follows the law

$$\sigma \simeq A_0 \exp\left(-E_g/2KT\right), \qquad (1.1)$$

where K is the Boltzmann constant and A_0 is a constant characteristic for a given material.

Since the density of free electrons in a metal remains constant, the electrical conductivity of metals is determined by the temperature dependence of the mobility of electrons, and falls slowly with increasing temperature.

If we take the logarithm of both sides of (1.1), we find that

$$\ln \sigma = \ln A_0 - E_g/2KT. \tag{1.2}$$

The graph of the left-hand side of this equation against the reciprocal of temperature is a straight line whose slope φ is an important parameter of the semiconductor because it governs its electrical and optical properties, including the gap width E_g. The slope is given by

$$\tan \varphi = E/2K.$$

It is important to note that the logarithm of the conductivity as a function of 1/T is an inclined straight line only in the case of pure semiconductors, free of extraneous impurities and commonly referred to as *intrinsic* semiconductors. Dopants introduced into semiconductors, usually to ensure that they have n- or p-type conductivity, are found to occupy energy levels in the forbidden band near the bottom of the conduction band (donors), and readily release electrons into the conduction band under slight thermal or optical excitation, or they occupy levels near the top of the conduction band (acceptors), which readily collect electrons from the filled bottom band, so that the crystal acquires pure p-type conductivity without any motion electrons in the upper band.

In these extrinsic (impurity) semiconductors, the dependence of ln σ on 1/T is more complicated and consists of two inclined straight lines joined by a horizontal section. The slope of the straight line lying in the low-temperature region can be used to determine the activation energy, or the energy position of impurity levels in the forbidden band. The slope of the other straight line, lying in the high-temperature region, can be used to find the width E_g of the forbidden band (the band gap) in the extrinsic semiconductor. The horizontal segment is due to the constant density of electrons in the conduction band (donor levels have become exhausted) in the particular interval of intermediate temperatures.

The nature of the temperature dependence of the properties of semiconductors cannot be used as a basis for a clear definition of semiconductors and of the way they differ from other media [11]. Numerous cases of departure from the above relationships have been investigated both theoretically and experimentally. Thus, in highly doped semiconductors, the electrical conductivity does not increase with increasing temperature but, instead, falls slightly, almost as in metals. Semiconducting superconductors have been discovered, in which the electrical conductivity rises sharply as the temperature approaches absolute zero. An extensive class of semiconductors has been found in which carrier transport is accomplished by ions rather than electrons, and the electrical conductivity exhibits completely different behavior (glassy semiconductors are examples of this). It is, therefore, more correct to define semiconductors as a class of materials whose properties can vary between very wide limits under the influence of external factors (temperature, illumination, pressure, electrical and magnetic fields, and so on). It is precisely this feature of semiconductors that has led to the development of exceptionally sensitive photo- and thermoresistors, electronic devices such as diodes, transistors, thyristors, electric and magnetic field detectors, particle detectors, strain gauges, and so on, which could not have been made from metals or dielectrics.

1.2 Absorption of light by semiconductors

In quantum mechanics, all elementary particles, including electrons, are also assigned wave properties. The motion of elementary particles is then described in terms of not only quantities such as the energy E and momentum p, but also, for example, the wavelength λ, the frequency v, and the wave vector $\mathbf{k} = p/h$ where h is Planck's constant. We then have $E = hv$ and $p = h/\lambda$. The band structure of a crystal is usually plotted in the form of the $E-\mathbf{k}$ diagram on which the energies are given in electron volts (eV) and the wave vector $\mathbf{k}$ in units of the lattice constant of the crystal. Arrows on the $\mathbf{k}$ axis are used to indicate the crystallographic orientation and hence show the variation in the band structure throughout the crystal.

The form of the $E-\mathbf{k}$ relation enables us to understand the nature of interband transitions in a given semiconductor and, in particular, to determine whether they are "direct" or "indirect"

[10]. When the minimum of the conduction band and the maximum of the valence band coincide in k-space, the transfer of an electron across the forbidden band as a result of optical or thermal excitation occurs vertically, wihout change in the wave vector. These are the so-called *direct transitions.* If, on the other hand, the band extrema do not coincide and k must change, the transfer of the electron is mediated by, for example, a quantum of lattice oscillations (the *phonon*), or an impurity ion, which gives up a fraction of its energy to the electron. Such transitions are referred to as *indirect.*

The structure of the energy bands of many semiconductors is quite complicated, so that the absorption of radiation is mixed in character: absorption begins with indirect transitions (the transfer of an electron then requires less energy, and low-energy photons are absorbed first) and the process continues at high energies exclusively via direct transitions.

It is also important to note that the width of the forbidden band, determined from the measured optical absorption, is very dependent on free-carrier density in the semiconductor, the temperature, and the presence of impurity levels in the forbidden band. When states near the bottom of the conduction band and the top of the valence band are filled with carriers, optical measurements yield a higher value for the gap width of an extrinsic semiconductor than for a pure intrinsic semiconductor. When the impurity band overlaps the edge of the nearest allowed band, which may occur in a highly doped semiconductor, the gap width is reduced, and this has an appreciable effect on the position of the *fundamental absorption edge* (which becomes shifted toward longer wavelength).

The absorption coefficient α of a semiconductor is defined by the equation

$$N_l = N_0 \exp(-\alpha l), \qquad (1.3)$$

where N_i is the flux of photons reaching the depth l in the semiconductor and N_0 is the photon flux density entering the semiconductor (some of the incident radiation is reflected by the surface). This means that the wave energy is reduced by a factor of e over a distance equal to $1/\alpha$.

In the spectral region in which αl is relatively small and some of the incident light passes through the semiconductor, the resultant transmission coefficient T must be calculated with allowance for multiple reflections in the interior of the semiconducting plate, using the formula [25]

$$T = \frac{(1-r)^2 \exp(-\alpha l)}{1 - r^2 \exp(-2\alpha l)} \, ,$$

where the reflection coefficient of the semiconductor surface (assuming that the front and rear surfaces are optically identical) is given by

$$r = \frac{(n-1)^2 + k^2}{(n+1)^2 + k^2} \, , \tag{1.4}$$

where n and k are, respectively, the refractive index and the absorption coefficient of the semiconductor.

The resultant reflection coefficient R [23] of a plate of this kind (with allowance for multiple reflections inside the plate) can be calculated from the formula

$$R = r_0 + \frac{(1-r_0)^2 r_2 \exp(-2\alpha l)}{1 - r_0 r_2 \exp(-2\alpha l)} \, ,$$

where r_0 is the single reflection coefficient of the outer surface (air-semiconductor boundary) and r_2 is the single reflection coefficient of the back surface (semiconductor-air boundary).

The absorption coefficient α of the material is related to its absorption index k by [25]

$$\alpha = 4\pi k/\lambda. \tag{1.5}$$

Thus, the values of k and α for a given material can be determined by measuring the intensity of optical radiation transmitted by semiconducting specimens of different but accurately measured thickness. When the results of such measurements are analyzed, it must be remembered that the radiation is reflected from both surfaces of the plate and that optical losses occur on both the front and the rear surfaces. Correct measurements, which result in a considerable simplification of calculations and much higher accuracy, require the use of specimens of different thickness but identical and carefully prepared surfaces (so that the reflection and transmission coefficients are the same). Interference effects must be absent. In the spectral region in which k is small, the refractive index n can be found from the Fresnel formulas, using the measured reflection coefficient of one of the surfaces of the semiconducting plate forming the boundary with the ambient air. Reflection from the second (rear) surface is best eliminated in some way, for example, by coarse grinding or by suitably slanting the polished back surface.

Solar cells are usually in the form of a two-layer semiconducting wafer: a highly doped silicon layer is formed on a relatively thick weakly doped base (this is the simplest model of a solar cell).

Calculations of the optical properties of such two-layer structures [23] involve averaging the reflection coefficient over the phase angle or, in other words, the phase thickness of the layers (for layer thickness much greater than the wavelength, the phase thickness becomes inadmissibly large, and very dependent on small changes in the layer thickness and wavelength, which is not consistent with the true physical situation). The formula for the transmission coefficient of this type of two-layer structure has the following form after the averaging has been carried out:

$$T = \frac{|g_0|^2 |t_1|^2 \exp(-k_d l_d 4\pi/\lambda)}{1 - |f_0|^2 |r_1|^2 \exp(-k_d l_d 8\pi/\lambda)} ,$$

where g_0 and f_0 are the Fresnel coefficients of the boundary between the highly doped silicon and the ambient air, t_1 and r_1 are the amplitude transmission and reflection coefficients of the boundary between the highly doped silicon and the silicon base layer, and k_d and l_d are, respectively, the absorption index and the thickness of the highly doped silicon layer.

If the optical constants of the two silicon layers are known, and if we measure the optical transmission of the two-layer silicon sandwich, we can determine the thickness of the weakly doped layer (of course, on the assumption that the two silicon layers are doped uniformly). The transmission coefficient of the two-layer structure for a transparent base layer (which is the situation in the long-wave spectral region beyond the fundamental absorption band of silicon) is found to depend on the thickness of the highly doped layer (see [23], Fig. 1.5).

The optical constants of the highly doped layer of silicon (and of other semiconducting materials) are, methodologically, the most difficult to determine, but this can be carried out by the method proposed in [26]. The practical implementation of this method

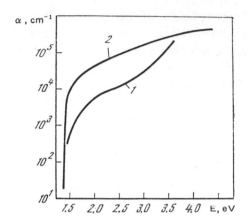

FIG. 1.1. Absorption coefficient of single-crystal silicon (1) and gallium arsenide (2) as a function of the incident photon energy at room temperature.

requires the availability of a material with known free-carrier concentration and a measurement of the spectral dependence of the transmission coefficient. For silicon with carrier concentration of between $2.2 \cdot 10^{18}$ and $11.5 \cdot 10^{19}$ cm^{-3} such calibrating curves are given in [26]. Subsequent calculations can be performed if the real part ϵ_0 of the permittivity ϵ_1 is known. For pure intrinsic silicon, $\epsilon_0 = 11.67$ [27]. The permittivity ϵ_1 is determined from the wavelength dependence of the reflection coefficient $R(\lambda)$ by graphical extrapolation, and the quantities n and k are then calculated [26].

Optical methods are thus seen to enable us to determine the geometrical thickness of the semiconducting layers taking part in the absorption process, the optical constants n and k, and their wavelength dependence, as well as the wavelength dependence of the absorption coefficient α of the semiconducting material [see (1.5)].

Figure 1.1 shows the absorption coefficient α as a function of the energy of incident photons for silicon [28] and gallium arsenide [29,30], the two materials most frequently used in semiconductor photoelectric energy conversion. It is precisely these materials that have been used to develop the most efficient solar cells. In Fig. 1.1, as elsewhere in this book, the results of measurements are given for single-crystal semiconducting materials at room temperature, unless stated to the contrary.

The obvious differences between the wavelength dependence of the absorption coefficient of the above two semiconducting materials are due to differences between their band structures and the nature of optical transitions. In gallium arsenide, the band-band transitions are direct optical transitions, so that absorption rises sharply as soon as photons with energies exceeding the band gap of gallium arsenide appear, and the absorption coefficient α rapidly reaches 10^4-10^5 cm^{-1}. Phonon-assisted absorption in silicon (in the fundamental band) begins with indirect transitions at 1.1 eV, and the absorption coefficient rises relatively slowly (up to 10^3-10^4 cm^{-1}). It is only when the photon energy reaches about 2.5 eV that the band-band transitions become direct, and there is a sharp rise in absorption.

Germanium is also characterized by indirect optical transitions which begin at 0.62 eV (absorption coefficient between 1 and 100 cm^{-1}), and it is only for photon energies in excess of 0.81 eV that fundamental absorption is determined by direct transitions [27].

It is important to note that the thermal width of the band gap of a semiconductor, calculated from the above temperature dependence of electrical conductivity, is usually the same as the optical gap width, determined by the fundamental absorption edge which coincides with the beginning of indirect band-band transitions.

The wavelength dependence of the absorption coefficient (Fig. 1.1) shows that, when silicon is used, a large proportion of the incident solar radiation (radiation below 1.1 μm) can be transformed into electric current, i.e., more than 74% of the extra-atmospheric solar radiation (AMO radiation) can be so converted. In the case of gallium arsenide, the photoactive radiation (i.e., radiation capable of taking electrons across the band gap) contains wavelengths below 0.9 μm, and this restricts the conversion process to only 63% of the AMO solar energy. However, indirect optical transitions and low values of the absorption coefficient near the fundamental absorption edge ensure that the thickness of silicon that will absorb the entire photoactive radiation must be at least 250 μm, whereas the corresponding value for the gallium arsenide cell is 2-5 μm. This feature of the wavelength dependence of absorption must be borne in mind when highly efficient and inexpensive thin-film solar cells are designed.

When the energy of the incident photons is so low that they

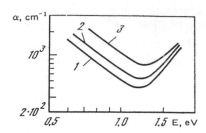

FIG. 1.2. Absorption coefficient of boron-doped silicon as a function of photon energy at 300 K: 1-3) boron concentrations of $6 \cdot 10^{19}$, $9.7 \cdot 10^{19}$, and $2 \cdot 10^{20}$ cm^{-3}, respectively.

cannot transfer electrons from the valence band into the conduction band, the effect of the incident radiation may be to force the electrons into transitions within allowed bands. This affects the absorption spectrum in the long-wave region immediately beyond the fundamental absorption edge, and is referred to as free-carrier absorption which increases with the concentration of ionized impurities and, consequently, with free-carrier concentration in the semiconductor. It also increases with the number of free carriers injected into the semiconductor under the influence of light or electric current.

Studies of the absorption spectra of long-wavelength radiation in semiconductors have led to the discovery of several further characteristic types of absorption, namely, absorption by lattice vibrations, impurity absorption, and excitonic absorption involving the excitation of an electron-hole pair but having no influence on free-carrier concentration because it is an excited state (and not the electron and the hole separately) that then moves through the crystal.

Figure 1.2 shows the experimental curves obtained for doped silicon. They confirm that the infrared absorption coefficient is a function of free-carrier concentration [11].

Absorption spectra provide us with extensive and useful information about the structural features of the crystal (including the dopant concentration). They enable us to determine the impurity activation energy and, hence, the position of the energy levels occupied by impurities in the band gap. Absorption spectra can even be used to investigate such fine effects as the presence of dissolved oxygen in silicon (by examining the characteristic absorption band at 9 μm) and to determine the concentration of oxygen in silicon [31].

1.3 Reflection of optical radiation from the surface of a semiconductor

While reflection by semiconductors in the region of the fundamental absorption edge is practically independent of the concentration of impurities (which ionize at room temperature), in the long-wavelength region there is a sharp rise in the reflection coefficient with increasing impurity concentration and, hence, increasing carrier concentration in the semiconductor.

Infrared reflection

The interaction of radiation with free carriers can be examined in terms of the classical electromagnetic theory of radiation [25,32], whose predictions are very similar to those derived from the quantum-mechanical theory of dispersion [33].

Theory and experiment. The permittivity of a semiconductor is equal to the square of its complex refractive index, i.e.,

$$\varepsilon_1 = (n-ik)^2, \tag{1.6}$$

and is determined by the concentration N and effective mass m* of carriers, the electron charge q, and the angular frequency ω of the incident radiation [25].

Equating the real and imaginary parts in (1.6), and using the relationship between the optical and electrical properties of crystals, we obtain

$$n^2-k^2=\varepsilon_1, \quad 2nk=4\pi\sigma/\omega, \tag{1.7}$$

where σ is the conductivity, n is the refractive index, and k is the absorption index of the semiconductor.

Analysis of the process of electrical conductivity in a high-frequency field [32] leads to the following expressions for the conductivity σ and permittivity ϵ_1:

$$\sigma = Nq^2/m^* < \tau/(1+\omega^2\tau^2) >, \qquad (1.8)$$

$$\varepsilon_1 = \varepsilon_0 - 4\pi\chi_c, \qquad (1.9)$$

where the polarizability averaged over all relaxation times τ and free-carrier energies is given by

$$\chi_c = -Nq^2/m^* < \tau^2/(1+\omega^2\tau^2) >. \qquad (1.10)$$

When the frequency of the incident radiation is much greater than $1/\tau$ and $\omega\tau \gg 1$, the polarizability χ_c is independent of τ and is given by

$$\chi_c \approx -Nq^2/m^*\omega^2. \qquad (1.11)$$

The relationships given by (1.6)-(1.11) were the first to be used in the determination of the effective mass of carriers by optical methods [34]. The carrier concentration was determined with the aid of the Hall effect. Use was also made of the wavelength dependence of the infrared absorption and refractive indices of highly doped germanium. The resulting data show that when the reflection coefficient r is calculated [see formula (1.4)], the term k^2 can be neglected up to wavelengths of the order of 15 μm. This is also valid for the other semiconductors used in solar cells.

The polarizability χ_c given by (1.11) increases with increasing wavelength whereas the permittivity decreases with increasing wavelength in accordance with (1.9). The permittivity may tend to zero when χ_c is large enough. The frequency at which this phenomenon sets in is referred to as the plasma frequency ω_p, which can be calculated from the following condition [32]:

$$\varepsilon_0 = 4\pi|\chi_c| = 4\pi Nq^2/m^*\omega_p^2,$$

which readily yields

$$\omega_p = (4\pi Nq^2/m^*\varepsilon_0)^{1/2}$$

and, consequently,

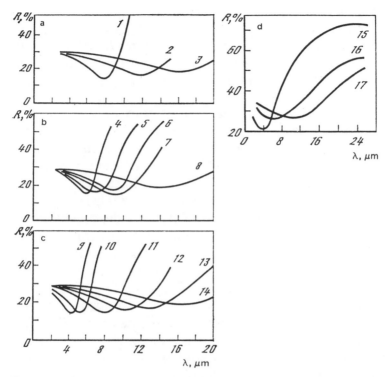

FIG. 1.3. Wavelength dependence of the reflection coefficient of silicon doped with antimony (a), arsenic (b), and phosphorus (c), and of gallium arsenide doped with zinc (d) for different concentrations of electrons (a-c) and holes (d) as carriers: l) $4.47 \cdot 10^{19}$ cm^{-3}, 2) $1.66 \cdot 10^{19}$, 3) $8.38 \cdot 10^{18}$, 4) $9.03 \cdot 10^{19}$, 5) $5.05 \cdot 10^{19}$, 6) $3.48 \cdot 10^{19}$, 7) $2.84 \cdot 10^{19}$, 8) $8.77 \cdot 10^{18}$, 9) $1.67 \cdot 10^{20}$, 10) $1.02 \cdot 10^{20}$, 11) $4.38 \cdot 10^{19}$, 12) $2.05 \cdot 10^{19}$, 13) $1.27 \cdot 10^{19}$, 14) $7.4 \cdot 10^{18}$, 15) $1.5 \cdot 10^{20}$, 16) $3.2 \cdot 10^{19}$, 17) $1.7 \cdot 10^{19}$ cm^{-3}.

$$\lambda_p = (\pi \varepsilon_0 c^2 m^* / N q^2)^{1/2},$$

where c is the velocity of light.

The phenomenon of plasma resonance occurs at wavelengths for which k^2 is small, and n is also reduced since its variation follows the wavelength dependence of ϵ_1 in accordance with (1.7). The single-reflection coefficient r given by (1.4) reaches its minimum value r_{min} for a low value of k and $n \cong 1$, and this corresponds to the plasma resonance region on the spectral reflection curve.

Figure 1.3 shows the wavelength dependence of the reflection coefficient of silicon doped with antimony, arsenic, and phosphorus for different values of the concentration (from $7.4 \cdot 10^{18}$ to $1.67 \cdot 10^{20}$ cm^{-3}) of free carriers (electrons) [35], and also for gallium arsenide doped with zinc for different free-carrier (hole) concentrations (from $3.2 \cdot 10^{19}$ to $1.5 \cdot 10^{20}$ cm^{-3}) [36]. The curves clearly show the position of the plasma resonance minimum and the dependence of the minimum-reflection wavelength on free-carrier concentration.

The value of r_{min} is determined by the absorption index k because, when $n \cong 1$, the reflection coefficient r decreases with decreasing k [as can be seen from (1.4)]. In its turn, the absorption index depends on the relaxation time because it affects conductivity. On the other hand, the polarizability is independent of τ and is determined by the band structure and free-carrier concentration.

The absorption of light by free carriers increases with increasing wavelength, and the rise in k leads to a rise in the reflection coefficient. The wavelength dependence of the reflection coefficient of doped semiconductors must, therefore, pass through a minimum, and this has indeed been confirmed experimentally (see Fig. 1.3). The polarizability is proportional to the product $N\lambda^2$. As the free-carrier concentration N increases, the same value of polarizability (in particular, $|\chi_c|$ for which $n \cong 1$) can be attained for lower wavelength λ. It is precisely for this reason that an increase in the free-carrier concentration is accompanied by a shift in the position of r_{min} toward shorter wavelengths, and the value of r_{min} itself decreases at the same time because k decreases.

This property of the infrared reflection spectra of doped semiconductors can be used as a basis for an optical method of determining the carrier concentration N from reflection spectra. The position of the plasma minimum for highly doped n- and p-type silicon ($\lambda_{n\text{-}Si}$ and $\lambda_{p\text{-}Si}$) is given by the following expressions (Fig. 1.4) [37]:

$$N_p = 3{,}27 \cdot 10^{21} \lambda_{p\text{-}Si}^{-2,11}, \quad N_n = 6{,}29 \cdot 10^{21} \lambda_{n\text{-}Si}^{-2,42},$$

which provide a good approximation to the experimental data. The carrier concentration in standard specimens was determined by measuring the layer resistance by the four-probe method, using Irvin's curves for the resistivity of p- and n-type silicon as a function of free-carrier concentration [31,38]. The disadvantage of this

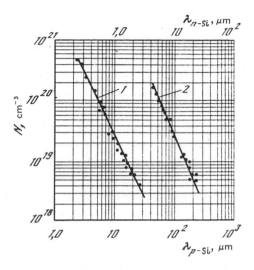

FIG. 1.4. Carrier concentration as a function of wavelength at the plasma minimum for p-Si (1) and n-Si (2). The points are experimental.

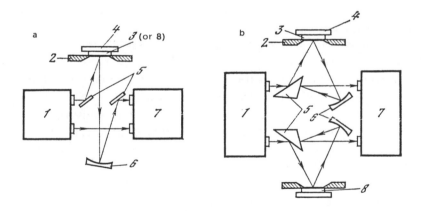

FIG. 1.5. Ray paths in the IKS-14 (a) and Hitachi (b) spectrophotometers, modified for reflection measurements: 1) source; 2) specimen holder; 3) specimen; 4) small thermostat; 5) flat mirrors; 6) concave mirror; 7) monochromator and detector of radiation; 8) reference aluminum mirror.

method is the relatively poor accuracy with which the position of the plasma minimum can be located in the case of weakly doped specimens.

More complicated but more accurate methods of determining concentration, mobility, and effective mass of free carriers from the reflection spectra of doped semiconductors are described in [39-42]. In some of these papers, the measured reflection curves are compared with the standard curves between 1 and 25 μm.

By studying the reflection spectra of the surfaces of semiconductors it has been possible to obtain information not only on the electrophysical properties of crystals but also on the state of these surfaces, i.e., the quality of chemical and mechanical surface treatment. This can be done despite the fact that it is difficult to record the resultant and diffuse reflection in the infrared [43] (usually only the specular component of the reflection coefficient and its temperature dependence are measured [44]).

Method of measurement. The reflection coefficient can be measured with an infrared spectrophotometer, for example, the Soviet instrument IKS-14 (0.7-25 μm) or the Japanese Hitachi instrument (2-50 μm). Both spectrophotometers essentially measure the transmission coefficient, and special devices [41,45] have to be inserted between the source and the receiver (Fig. 1.5) to determine the reflection coefficient.

A convenient specimen holder and a small thermostat suitable for such measurements are described in [44]. The circuit controlling the thermostat is based on the relay principle and can maintain any given temperature in the range 30-300°C to within ±2°C.

Measurements of the background thermal emission by the specimens under investigation have shown that, even in the far infrared, the background radiation at 200°C does not exceed 14% of the reflected flux [44]. The background emission by the heated reference mirror amounts to less than 1% because of the high degree of blackness of the aluminum films [46].

Quality of mechanical treatment applied to the semiconductor surface. Measurements of the infrared reflection spectra at room temperature, performed with infrared spectrophotometers, can be used to estimate the quality of the semiconductor surface after mechanical treatment, for example, using the procedure employed for n-type gallium arsenide [47] (of course, successful surface moni-

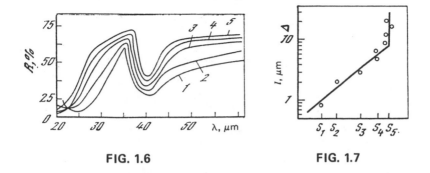

FIG. 1.6   FIG. 1.7

FIG. 1.6. Reflection coefficient of a mechanically polished surface of n-type gallium arsenide in the infrared before (1) and after (2-5) the outermost layer has been removed by etching. Thickness removed: 2) $0.9\,\mu m$; 3) 2.2; 4) 5.8; 5) 13, 18, 25 μm.

FIG. 1.7. Thickness of layer on gallium arsenide, removed by etching, as a function of the area bounded by the spectral reflection coefficient curve (the measurements were made after each stage of etching): S_1-S_5) areas bounded by curves 1-5 of Fig. 1.6.

toring requires that the wavelength of the incident radiation must be comparable with the depth of surface defects). The procedure was used to examine relatively coarse surface damage produced by mechanical treatment with an abrasive with grain size of about 10 μm. The surface was examined in the spectral region between 20 and 60 μm.

In this range, the reflection curve obtained for doped gallium arsenide (Fig. 1.6) has two characteristic minima, namely, one at 20-25 μm (due to plasma resonance between the incident radiation and free carriers) and the other at 37-42 μm (due to absorption as a result of interaction with thermal lattice vibrations). Removal of a surface layer of a given thickness had no effect on the wavelength dependence of the reflection coefficient, which was found to be the same as that obtained for undamaged n-type gallium arsenide with the same impurity concentration. The damaged layer was removed with an etching solution consisting of a mixture of sulfuric acid, hydrogen peroxide, and water. The thickness of the layer removed in this way was determined by weighing the specimens before and after etching. The depth of the damaged surface layer can be determined with adequate accuracy by measuring the area bounded by the reflection coefficient curve as a function of the thickness of the

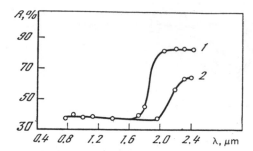

FIG. 1.8. Wavelength dependence of the reflection coefficient of the surface of a polycrystalline germanium wafer (resistivity 60 $\Omega \cdot$cm) coated on the rear face with aluminum. Temperatures: 1, 2) 20 and 200°C, respectively.

removed layer (Fig. 1.7). The area under the reflection-coefficient curve ceases to increase after 8.8 μm of the damaged layer has been removed. The depth of damage is thus about 0.9 of the average diameter of the abrasive grain used in mechanical grinding of the gallium arsenide surface.

Temperature dependence of the band gap. A small thermostat [44] can be used to investigate the specular reflection coefficient as a function of temperature, and to estimate the temperature variation of the semiconductor band. The change in reflection from the outer surface of a polycrystalline p-type germanium wafer (resistivity 60 $\Omega \cdot$cm) and a reflecting layer of aluminum deposited in vacuum on its back surface (Fig. 1.8) [44] was due to the temperature shift of the fundamental absorption edge of germanium.

The properties of thin highly doped semiconductor layers cannot be examined quantitatively by measuring the infrared reflection spectra when the depth of penetration of the radiation into the semiconducting material exceeds the layer thickness. This is confirmed by studies of thin p—n junctions produced by bombarding silicon with phosphorus ions [48]. The methods used in these experiments are an example of the composite approach to the study of the properties of semiconducting structures, in which a determination was made of the distribution of free-carrier concentration, the depth of the p—n junction, and the variation in these parameters during isothermal annealing.

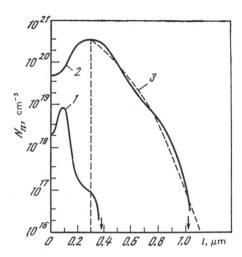

FIG. 1.9. Free-carrier concentration in the doped surface layer of silicon, produced by bombarding p-type silicon with phosphorus ions, as a function of distance from the surface: 1) before annealing, phosphorus ion dose $6 \cdot 10^3$ $\mu C/cm^2$; 2) same specimen after annealing for 2 hr at $850°C$; 3) theoretical curve corresponding to the diffusion of an impurity from an infinite source at 0.3 μm from the surface. Arrows show the position of the p–n junction.

Studies of thin doped layers on solar cells. In these experiments, p-type silicon with resistivity of 1 $\Omega \cdot cm$ was examined. The carefully polished surface was oriented along {111}, and was exposed to a separated beam of phosphorus ions with energies of 30 keV. The dose was 6000 $\mu C/cm^2$. The distribution of free-carrier concentration was investigated by successively removing silicon layers of 160-500 Å by anode oxidation in an 0.04N solution of potassium nitrate in ethylene glycole [49]. The conductivity of the layers removed in this way was measured by the four-probe method. The mean concentration of carriers in the removed layer was calculated from the data reported in [38,50]. The total depth of the doped layer was estimated by the grooving method [51].

The transmission and reflection coefficients were measured in the spectral range 1-25 μm, using the IKS-14 spectrophotometer and a special device to determine the specular reflection coefficient (see Fig. 1.5a).

Electrical measurements. Figure 1.9 (curve 1) shows the distribution of free-carrier concentration with depth in the doped layer. A relatively wide region with a negative concentration gradient is formed near the surface, and the maximum extends to a depth of about 0.12 μm. Thereafter, the concentration falls to the value corresponding to the original silicon. The shape of the curve is explained by the particular distribution of implanted-atom concentration [52] and by radiation defects: the concentration maximum is shifted toward the center [53-55]. Electron micrography of the silicon surface bombarded with phosphorus ions has demonstrated the amorphization of silicon down to a depth of 0.2 μm. The top 0.05-μm layer was found to have been converted from the single-crystal to the amorphous state [48]. Quantitative estimates of the carrier concentration within the layer (see curve 1 in Fig. 1.9) yield a figure of about 10^{18} cm^{-3} which is lower by a factor of about 1000 than the mean concentration of implanted phosphorus atoms ($3 \cdot 10^{21}$ cm^{-3}).

It is well known that a reduction in the number of radiation defects and an increase in the concentration of electrically active implanted phosphorus atoms can readily be achieved by thermal annealing of the specimens (see Fig. 1.9, curve 2). The depth of the p—n junction is then increased to 1 μm. The portion of curve 2 lying between 0.3 and 1 μm can be satisfactorily described in terms of the diffusion of the impurity from an infinite source into a semi-infinite body [56]. Integration of curve 2 shows that 4.2% of the phosphorus has diffused into the silicon from an initial layer 0.3 μm thick. The free-carrier concentration is found to decrease as the surface is approached, both before and after annealing. Electron micrographs recorded after each successive layer was removed showed that there was some residual damage of the single-crystal structure down to a depth of 0.15 μm. This is an indication of incomplete annealing, so that high free-carrier concentration could not be achieved in the layer.

Optical measurements. An attempt was made to investigate the distribution of carriers in the doped layer by determining the variation of the infrared reflection coefficient. The reflection coefficient of unannealed specimens after ion bombardment was found to be equal to that for undoped silicon. In this case, the surface carrier concentration was N $\cong 10^{18}$ cm^{-3}, but the depth of the junction was

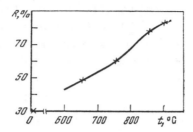

FIG. 1.10. Reflection coefficient of silicon, doped by ion bombardment with phosphorus ions, as a function of temperature. The coefficient was measured at the wavelength of 19 μm.

so small (0.2-0.3 μm) that the doped layer was highly transparent near λ = 19 μm. This suggests that, in shallow p—n junctions (depth <1 μm), the variation in the reflection coefficient as the successive silicon layers are removed does not reproduce the true carrier distribution in the doped layer because the reflection coefficient is determined not only by the surface layers, but also by all the underlying layers with lower free-carrier concentration.

This result was deduced from experimental data, but is also confirmed by calculation. Thus, substituting λ = 19 μm and k = 4.1 in (1.5) in the case of doped silicon [57], we find that the depth of penetration of light at which the radiation flux has been reduced by a factor of e is $1/\alpha$ = 0.4 μm. Since in these experiments the carrier concentration in well annealed doped layers was $3 \cdot 10^{20}$ cm^{-3} [48], whereas the value k = 4.1 reported in [57] corresponds to the concentration of $2.9 \cdot 10^{19}$ cm^{-3}, the agreement between the calculated and measured infrared transparency of the doped layer as a function of its thickness must be regarded as very satisfactory.

Nevertheless, these measurements of the reflection coefficient are useful. For example, they show that, for the same junction depth, the free-carrier concentration after thermal annealing is higher when the silicon dioxide film preventing phosphorus exodiffusion is present.

Moreover, the value of the infrared reflection coefficient can be used to judge the degree of annealing under given conditions. Figure 1.10 shows the reflection coefficient at λ = 19 μm for a silicon surface after bombardment with a phosphorus-ion beam (under the above conditions), followed by annealing for two hours, with the temperature rising from 600° to 890°C. The higher the free-carrier concentration in the surface layers of the doped region, the greater the reflection coefficient: R = 30% for $N_p = 10^{16}$ cm^{-3} and R = 84% for $N_n = 2.5 \cdot 10^{20}$ cm^{-3}.

Reflection in the ultraviolet region

Optical studies of thin doped layers of silicon and other semi-conducting materials are much easier and more accurate when they are performed with radiation that is strongly absorbed by the semi-conducting material. For example, ultraviolet radiation with wave-length in the range 0.2-0.4 μm is almost completely absorbed by a silicon layer only 0.05-0.1 μm thick. However, its optical properties in the short-wave part of the spectrum are practically independent (within wide limits) of the concentration of free carriers [58]. Ultraviolet reflection spectra have in fact helped us to establish the band structure of semiconductors. The reflection peaks that are characteristic for many semiconductors are explained by the sharp rise in the refractive index due to interband transitions when the band gap is large in those regions of the dependence of E on k in which $\mathbf{k} \neq 0$ [10,11,25].

The ultraviolet reflection spectra have also been used in accurate monitoring of the quality of mechanical and chemical polishing of the surface of semiconducting crystals. This is clearly demonstrated by Figs. 1.11 and 1.12 which show the wavelength dependence of the reflection coefficient on the depth of defects remaining on the surface of silicon and gallium arsenide after polishing. It is only after the depth of defects due to additional polishing becomes less than the wavelength of ultraviolet (0.2-0.4 μm) and visible (0.4-0.75 μm) radiation used in such measurements that the reflection coefficient in these spectral bands ceases to vary (see Fig. 1.12).

Effective optical monitoring is facilitated by the fact that silicon and gallium arsenide produce reflection peaks in the ultraviolet region. For example, the state of silicon surfaces is best monitored at the wavelength of 0.28 μm at which the reflection coefficient of highly polished silicon is up to 70%. The difference between the reflection coefficients of surfaces subjected to different treatment can be enhanced by using an instrument in which ultraviolet radia-

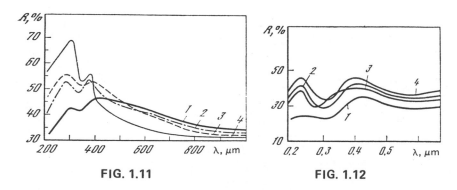

FIG. 1.11   FIG. 1.12

FIG. 1.11. Wavelength dependence of the reflection coefficient of the silicon surface after mechanical polishing with diamond paste and abrasive powder with different grain diameters: 1) 3 μm; 2) 1 μm; 3) 1 μm; 4) 0.1-02 μm; 1, 2, 4) polishing time 1 hr; 3) 2 hr.

FIG. 1.12. Wavelength dependence of the reflection coefficient of gallium arsenide after treatment with etching solution, with surface damage remaining at the following depths: 1) 1 μm; 2 and 3) 0.2-0.3 μm; 4) less than 0.2 μm.

tion is multiply reflected from a set of plates with identically treated surfaces (Fig. 1.13) [59]. Another set of highly polished plates is used in the same instrument to isolate the 0.28-μm radiation which is particularly suitable for monitoring the state of silicon surfaces.

More complex systems used to monitor the state of the surface, and capable of isolating radiation within the necessary wavelength interval, incorporate quartz prisms, diffraction gratings, or parabolic and rotating mirrors.

It is thus clear that ultraviolet and infrared reflection spectra provide information on the electrophysical and optical properties of crystals. In particular, they can be used to estimate the free-carrier concentration, the quality of the surface, the degree of annealing, and the parameters of band structure, including the width of the band gap of the semiconductor and its temperature dependence.

1.4 Conversion of optical radiation into electric power in semiconductor solar cells

Photocells based on the photoelectric effect in semiconducting structures with a barrier layer (photovoltaic effect) can directly con-

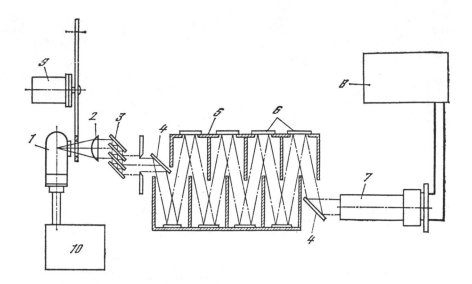

FIG. 1.13. System used for the optical monitoring of the depth of damage on the surface of semiconductors: 1) hydrogen lamp; 2) quartz lens; 3) filter consisting of silicon wafers treated with chromium oxide; 4) flat mirrors; 5) partitions; 6) wafers under investigation; 7) FEU-57 or FEU-39 quartz window photomultiplier; 8) photocurrent measuring unit; 9) chopper; 10) supplies for hydrogen lamp.

vert optical radiation incident upon them into electric power. They are thus electric power generators (in contrast to photoresistors and photocells exploiting the external photoelectric effect) and do not require an external voltage source.

Since the discovery, in the middle of the last century, of the photoelectric properties of selenium, and the development at the turn of the century of the first photoelectric converters of radiation into small electric signals (based on selenium and copper/copper oxide heterosystems), there have been frequent attempts aimed at increasing the efficiency of such converters and exploiting them as sources of useful electric power. Improved technology, and also the use of optical filters, resulted in selenium photocells whose spectral sensitivity was practically the same as that of the human eye. Improved selenium photocells found extensive applications as exposure meters in photography and cinephotography. However, the efficiency of such photocells did not exceed 0.5%. The successful application of the photoelectric method of energy conversion began only after the advent of the band theory of the electronic structure of semiconductors, the development of methods for their purification and controlled doping, and the elucidation of the dominant role of the barrier layer on the separation boundary between semiconductors with different types of conductivity. A brief report of the development of the silicon solar cell with an efficiency of about 6% appeared in 1954, and by 1958 both Soviet and American satellites carried silicon solar batteries supplying electric power for the electronics. Since then, the efficiency of solar cells has been substantially increased. This has been facilitated by the increasingly better understanding of physical phenomena occurring in solar cells, by the introduction of continuously improving fabrication techniques, and by the development of new and improved cell designs relying on a variety of semiconducting materials.

Most of the barrier-layer photocells developed during the first few decades of development of photoelectricity (selenium, gallium sulfide, silver sulfide, copper sulfide, germanium, and others) continue to be used mostly as radiation detectors. Silicon photocells and, recently, photocells made of gallium arsenide and other wide-gap semiconductors, have found extensive applications as photoelectric converters of solar radiation, or solar cells, because of their high efficiency (15-22% in the case of better samples; or even 27-28% when complex cascade systems are employed [60]).

Photoelectric generators for the direct conversion of solar radiation into electric power consist of a large number of series and parallel connected photoconverters, and are called solar batteries. Modern solar batteries generate considerable amounts of electric power and are used to supply radio circuits, communication systems, cosmic-ray counters, space probes, and many other automatic devices used on the ground.

Principle of the solar cell

The semiconductor solar cell, for example, the silicon solar cell, usually consists of two semiconducting layers with p- and n-type conductivity, respectively. The two layers are in intimate contact with one another, and the transition region (boundary) between them, which lies in the interior of the semiconducting material, is called the p—n junction (it was previously often referred to as the barrier layer).

In the state of equilibrium, the Fermi energy must be the same throughout the material [9-11]. This is so because of the presence of the charged double layer, called the space-charge layer, in the region of the p—n junction, and the associated electrostatic potential (Fig. 1.14). The height of the potential barrier is equal to the difference between the bottom of the conduction bands of the n- and p-type materials. It is important to note that the position of the Fermi level and, consequently, the height of the barrier, depend on the temperature and the concentration of impurities in the semi-conducting materials on both sides of the p—n junction which, on the one hand, enables us to vary the properties of the p—n junction within wide limits and, on the other hand, determines the relatively strong temperature dependence of its optical and photoelectric characteristics [5,6].

Optical radiation incident on the surface of the semiconductor structure containing the p—n junction produces electron-hole pairs (mostly near the surface). The concentration of these pairs decreases gradually with distance from the surface in the direction of the p—n junction. When the distance of the p—n junction from the surface is less than the depth $1/\alpha$ of penetration of light, the electron-hole pairs are produced after the p—n junction as well. If the distance between the junction and the point at which the pairs appear is less than the diffusion length, the pairs diffuse toward the p—n junction and separate under the influence of the junction field. The electrons enter the n-type part of the junction and the holes the p-type part. A potential difference thus appears between the external metal electrodes attached to the p- and n-regions, and a current flows through the load resistor [12-16].

The excess minority carriers diffusing toward the p—n junction become separated because of the presence of the potential barrier. The accumulation of the excess electrons (separated by the junction) in the n-type region and of holes in the p-type region of the photo-converter leads to the compensation of the space charge localized in the p—n junction, i.e., to the appearance of an electric field which is opposite in direction to the existing field.

Thus, the appearance of the potential difference between the external electrodes is accompanied under illumination by a change

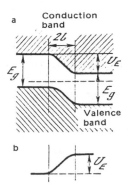

FIG. 1.14. Energy band structure of a semiconductor in the region of unilluminated p—n junction (a) and electrostatic potential distribution (b): 2l) width of space-charge region in the barrier layer; U_E) electrostatic potential on the boundary between the n- and p-type regions in equilibrium; E_g) width of band gap; dashed line) equilibrium position of Fermi level.

in the potential barrier in the previously unilluminated p—n junction. The resulting photo-emf reduces this barrier which, in turn, gives rise to the flow of opposing currents (in addition to those present in equilibrium) of electrons and holes from the n-type and p-type regions, respectively. These currents are practically equal to the current in the forward direction, which appears when a voltage is applied to the p—n junction. Thus, from the moment the illumination is turned on, the accumulation of excess (as compared with the equilibrium state) concentration of electrons and holes in the respective parts of the p—n junction is accompanied by a reduction in the barrier height or in the electrostatic potential U_E (see Fig. 1.14). This produces an increase in the current flowing through the external load and also an increase in the density of the electron and hole currents flowing in opposite directions through the p—n junction. The stationary state is established when the number of excess pairs produced by the illuminating light becomes equal to the number of pairs escaping through the p—n junction or into the external load. As a rule, this occurs in a few thousandths of a second after the illumination is turned on [17,18].

Quantum yield of internal photoelectric effect

Measurement of the short-circuit current of a solar cell simultaneously with the spectral composition and density of the optical radiation incident upon it enables us to deduce the efficiency of the different stages of conversion of radiation into electric power within the cell.

The first step is, of course, to define the flux of radiation involved in the particular process. When the short-circuit current of the solar cell is a linear function of the radiation flux density, we have

$$I_{sc2}(\lambda) = I_{sc1}(\lambda)/(1-r(\lambda)),$$

where $I_{sc2}(\lambda)$ and $I_{sc1}(\lambda)$ are, respectively, the short-circuit currents corresponding to absorbed and incident radiation, and $r(\lambda)$ is the single-reflection coefficient. All three quantities refer to a particular wavelength.

To find $r(\lambda)$, we must know n and k but, in the region of the fundamental absorption band, in which k is small, it is sufficient to know only the refractive index n. When we are dealing with a semiconductor that has not been extensively investigated, so that only the width of the forbidden band E_g is known, but the optical constants have not as yet been determined, we can use the empirical Moss rule [25] to calculate n:

$$E_g n^4 = 173.$$

The short-circuit current of a cell, calculated per absorbed photon, is a useful quantity when one examines the quality of a solar cell and its performance. It is usually referred to as the effective quantum yield of a solar cell, Q_{eff}, and is a function of wavelength. Thus,

$$Q_{eff} = I_{sc2}/N_0,$$

where N_0 is the number of photons incident per unit area on the semiconductor and I_{sc2} is the short-circuit current (electrons per second).

The effective quantum yield of a solar cell depends on two other parameters, namely,

$$Q_{eff} = \beta\gamma, \tag{1.12}$$

where β is the quantum yield of the internal photoelectric effect, defined as the number of electron-hole pairs produced by photo-ionization, in the interior of the semiconductor absorbed per photon, and γ is the carrier collection efficiency (or, in other words, the carrier separation factor) of the potential barrier in the p—n junction. The latter gives the fraction of pairs produced by optical radiation that participate in forming the short-circuit current when an external recording device is connected to the cell.

It is commonly considered that the quantum yield of the photo-electric effect is equal to unity when each absorbed photon produces one electron-hole pair. High-precision measurements of this quantum yield are reported in [61] for silicon and are examined in detail in [62]. A high-precision setup was used to measure simultaneously the short-circuit current of semiconducting crystals containing a p—n junction and the resultant (diffuse plus specular) reflection from the surface. The spectrum of the incident optical radiation was deter-mined with a mirror monochromator, the crystal temperature was established by a cryostat and an electric furnace, and the energy of the photons was determined with a calibrated thermopile. Visible and ultraviolet (down to 0.254 μm) radiation reflected from the crystal surface was recorded by a phosphor plus photomultiplier arrangement.

The quantum yield of the internal photoelectric effect was calculated from the formula

$$\beta = I_{sc_1} / (1-r)\, qN_0\gamma, \tag{1.13}$$

where q is the electron charge since I_{sc1} is in energy units.

These measurements were carried out on crystals containing a p—n junction, and the photoelectric effect was detected directly by measuring the current generated in the circuit without applying an external bias. The experimental conditions were such as to ensure that the carrier collection efficiency was $\gamma = 1$ (or, at least, was con-stant throughout the wavelength interval employed) and also that

(1.13) could be used in the calculation. Because of this, the crystal selected for the experiments had a long minority-carrier diffusion length L_d in the upper doped layer. The depth l_d of the p–n junction was small and the condition $L_d > l_d$ was satisfied. The experiments were confined to visible and ultraviolet radiation.

Analysis of the resulting experimental data showed that the photoionization quantum yield in silicon was unity ($\beta = 1$) in a wide range of incident-photon energy ($E_g < h\nu < 2E_g$). At the upper end of this range ($h\nu > 2E_g$, i.e., in the ultraviolet), the photoionization quantum yield β was found to rise sharply, which was probably due to ionization by collision, i.e., the production of secondary electron-hole pairs at the expense of the excess kinetic energy of the primary pairs.

Thus, it may be considered that the first interaction between optical radiation and the semiconductor (in the interior of the crystal) occurs practically without loss, i.e., with efficiency approaching 100% in a wide range of wavelengths. However, in most semiconductors used in solar cells, an increase in the photon energy is accompanied by losses due to the finite width of the band gap of the semiconducting material, despite the fact that the photoionization quantum yield is unity (and also for $\gamma > 1$ in the ultraviolet) [63]. The use of solar cells of more complicated structure, for example, the cascade system, or cells with a graded gap (wide gap at the surface and narrower gap in the interior of the material, so that the absorption coefficient is a function of wavelength), enables us to avoid such optical and energy losses and to increase the efficiency of conversion of solar radiation into electric power.

Carrier collection efficiency

Optical radiation of different wavelength penetrates to different depth (i.e., the penetration depth is a function of photon energy) and each wavelength produces its own spatial distribution of the resulting electron-hole pairs (see Figs. 1.1 and 1.2). The subsequent fate of the pairs created in this way depends on their diffusion length in the given semiconducting material. When this length is large enough, excess minority carriers produced by the incident light will diffuse (even without a dragging electric field) toward the p–n junction, and will be separated by the junction field. The ratio of the diffusion length L and the distance l between the p–n junction and the point at which the electron-hole pairs are produced plays a decisive role in the efficiency of this stage of conversion of optical radiation in the interior of the semiconductor.

Let us consider two extreme orientations of the p–n junction in the semiconducting crystal relative to the direction of incidence of the optical radiation, namely, the perpendicular position (Fig. 1.15a) and the parallel position (Fig. 1.15b). We shall suppose that, in the first case, light passes through the entire crystal and l is equal to the thickness of the semiconducting wafer whereas, in the second case, the entire surface of the wafer of width d is illuminated.

It is clear that the collection efficiency in the perpendicular and parallel positions of the p–n junction is given by the following expressions, respectively:

$$\gamma = (L_n + L_p)/l \quad \text{and} \quad \gamma = (L_n + L_p)/d.$$

At first sight, the parallel position would seem to be preferable, since the most important point for the complete collection and separation of carriers is the distribution of pairs in the direction perpendicular to the p–n junction: uniform carrier generation within the crystal gives rise to favorable conditions for their diffusion toward the p–n junction, and subsequent spatial separation. The multijunction matrix solar cells based on this orientation of the p–n junction consist of a large number of microcells, the planes of which are parallel relative to the incident solar radiation (they are arranged at a small angle to it), and do in fact have high carrier collection efficiency in the long-wave part of the spectrum. They generate a considerable photo-emf per unit illuminated area [64,65]. However, it has been established, both by calculation and experimentally, that because the microcells have very small dimensions, the recombination of light-generated pairs on the illuminated surface plays a much greater role in the parallel position of the p–n junction than in the perpendicular position. This means that the collection efficiency in the short-wave part of the spectrum can be increased by depositing an additional layer, doped with an impurity with the opposite conductivity type, on the surface facing the incident light, i.e., partial use has to be made of a structure with the perpendicular position of the p–n junction [66].

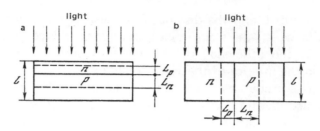

FIG. 1.15. Diagram illustrating the position of the p—n junction in the semiconducting crystal under perpendicular (a) and parallel (b) incidence of optical radiation. L_n, L_p) diffusion length of minority carriers in the n- and p-type regions, respectively; l) depth of penetration of light into the semiconductor.

While in the parallel position the concentration M of light-generated pairs decreases with distance from the surface of the semiconductor in both the n- and p-type regions, in the perpendicular position this is characteristic only for the region of the crystal facing the incident light, for example, the n-type region, whereas in the p-type region the highest number of pairs is generated at the p—n junction. The concentration of pairs at a depth l is obtained by the following expression, obtained by differentiating (1.3):

$$N = N_0 \alpha \exp(-\alpha l),$$

where N_0 is the number of photons incident per unit area of the semiconductor. The concentration of pairs, which decreases with depth in the semiconductor, can be calculated for the absorbing region from the function $\alpha(E)$ (see Fig. 1.1). Figure 1.16 shows the results of this calculation for silicon [15]. This figure clearly illustrates the carrier collection process for the perpendicular position of the p—n junction relative to the incident radiation (see Fig. 1.15a).

The ordinates of points on these curves are proportional to $\alpha \exp(-\alpha l)$ and the abscissas are proportional to the depth in the semiconductor from the illuminated surface. The area between the axes and each of the curves is proportional to the incident photon flux, and the area bounded by the curve and the ordinates corresponding to $l = l_d + L_n$ and $l = l_d - L_p$ (shaded region) is proportional to the short-circuit current of the silicon wafer containing the p—n junction.

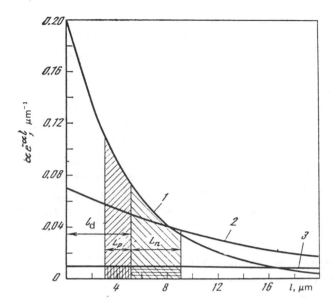

FIG. 1.16. Distribution of electron-hole pairs produced by optical radiation within the silicon for radiation of different wavelength incident at right angles to the plane of the p–n junction: 1) 0.619 μm, $\alpha = 2000$ cm^{-1}; 2) 0.81 μm, $\alpha = 700$ cm^{-1}; 3) 0.92 μm, $\alpha = 90$ cm^{-1}.

The ratio of the shaded area to the total area under the curve is thus a measure of the collection efficiency γ, in accordance with (1.13) (provided, of course, that the photoionization quantum yield is $\beta = 1$).

The planar design of the solar cells illustrated in Fig. 1.15a is the most widely used. Solar cells of this type have been constructed from a great variety of different materials. The optimum design of this type can readily be established by analyzing calculations similar to those performed for silicon and shown graphically in Fig. 1.16.

It is obvious that γ and I_{sc} can be increased by increasing the diffusion length of minority carriers on both sides of the p–n junction (L_n and L_p). This can be achieved by a suitable choice of the original materials in which high values of L must be maintained when the p–n junction is introduced. When the value of L in the portion of the semiconductor adjacent to the illuminated region (L_p in Fig. 1.15) cannot be increased, the p–n junction must be brought closer to the illuminated surface so as to satisfy the condition $L_p \gg l_d$ where l_d is the depth of the p—n junction, and all the carriers generated by the incident light can be collected and separated by the p—n junction field. Small p—n junction depths can be achieved by modern technologies [5,13,21].

A similar condition must be satisfied for the base region of the solar cell as well (it lies behind the p—n junction). The thickness of the solar cell, determined largely by the base region, must be less than the depth of penetration of the photoactive part of the incident radiation in the semiconductor (photon energy $h\nu > E_g$), and the diffusion length of minority carriers in the base region must correspond to the thickness of the cell and the depth of penetration of light.

1.5 Current-voltage characteristic and spectral sensitivity of a solar cell

The basic photoelectric properties of solar cells such as the current-voltage characteristic and spectral sensitivity are functions of both the optical and electrophysical properties of the semiconductor. Only a detailed analysis will reveal the factors that limit the efficiency of a given solar cell. However, this must be preceded by measurements of the basic parameters of the cell, which enable us to understand the reasons for the origin, nature, and predominant form of losses.

Current-voltage characteristic

Early theoretical and experimental work on the properties of solar cells showed that the current-voltage characteristic of a cell differs from that of the semiconductor diode by the presence of the term I_{ph} which represents the current generated by the incident light. Part of this current, I_D, flows through the diode and the other part, I, through the external load:

$$I_{ph} = I_D + I,$$ (1.14)

where

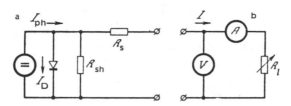

FIG. 1.17. Equivalent (a) and measuring (b) circuits of a solar cell.

$$I_D = I_0 (\exp (qU/KT) - 1) \tag{1.15}$$

is the usual dark characteristic, I_0 is the reverse saturation current of the p–n junction, q is the electron charge, T is the absolute temperature, K is the Boltzmann constant, and U is the voltage. When the cell is on open circuit, its resistance is infinite and I = 0. Equation (1.14) then yields the open circuit voltage across the solar cell:

$$U_{oc} = \ln (I_{ph}/I_0 + 1) KT/q.$$

A practical solar cell is characterized by the series resistance R_s, which consists of the series-connected resistances of the contacting layers (the resistances of the p- and n-type regions in the cell), the metal-semiconductor junction resistances, and the shunting resistance R_{sh}, which represents possible surface and volume leakage currents in parallel with the p–n junction. Allowance for these resistances and for recombination in the p–n junction leads to the following explicit expressions for the current-voltage characteristic, which includes the coefficient A:

$$\ln \left(\frac{I + I_{ph}}{I_0} - \frac{U - IR_s}{I_0 R_{sh}} + 1 \right) = \frac{q}{AKT} (U - IR_s). \tag{1.16}$$

This can be rewritten in the following more convenient form:

$$I = I_{ph} - I_0 \left(\exp \frac{q(U + IR_s)}{AKT} - 1 \right) - \frac{U + IR_s}{R_{sh}}, \tag{1.17}$$

which enables us to construct the equivalent and measuring circuits for the solar cell (Fig. 1.17).

Calculations of the current-voltage characteristics, based on (1.17) [67], have established the effect of the series and shunting resistances on the properties of solar cells. The results of these calculations are shown in Fig. 1.18. The output power P per unit area (1 cm^2) of a solar cell can be estimated from

$$P = (I_l U_l)_{\max} = \zeta I_{sc}\ U_{oc},$$

where ζ is the so-called curve factor, indicating the degree to which the characteristic approaches the square shape: $\zeta \cong 0.8$-0.9 means that the corresponding cells have a high output power. For modern silicon solar cells, the coefficient ζ is usually in the range 0.75-0.8. The shape of the current-voltage characteristic is relatively unaffected by a reduction in the shunting resistance from infinity to a value as low as 100 Ω (see Fig. 1.18). The output power of the solar cell is equally unaffected. On the other hand, a small change in the series resistance, for example, from 1 Ω to 5 Ω, leads to a sharp deterioration in the shape of the characteristic and to a considerable reduction in the output power.

The light and dark current-voltage characteristics of the solar cell were subsequently subjected to a still more detailed analysis. It was found that the reverse saturation current through the p—n junction depended on the applied voltage. As a rule, this current is the sum of two components. The equation for the current-voltage characteristic of a solar cell is therefore commonly written in the form [68]

$$I = I_{01}\left(\exp\left(\frac{q}{AKT}U\right) - 1\right) + I_{02}\left(\exp\left(\frac{q}{AKT}U\right) - 1\right) - I_{ph}.$$

where I_{01} is the reverse saturation current, determined by the diffusion mechanism of current flow through the thin p—n junction [69], and I_{02} is the reverse saturation current that appears after combination in the region of the p—n junction [70] (usually, $A = 2$ in this situation).

There are several relatively accurate methods for calculating I_0, R_s, R_{sh}, and the coefficient A from the measured light and dark current-voltage characteristics of solar cells [15,16,21,71]. Such data can be used to identify physical processes that restrict the efficiency of solar cells made from particular semiconducting materials.

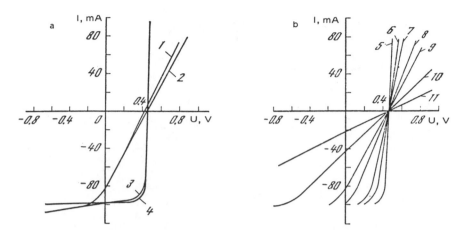

FIG. 1.18. Calculated current-voltage characteristics of solar cells with different combinations of series and shunt resistances (a) and different series resistances for an infinite shunting resistance (b) (I_{ph} = 0.1 A; I_0 = 10^{-9} A; q/KT = 40 V^{-1}): 1, 2) R_s = 5 Ω, R_{sh} = 100 and ∞, respectively; 3, 4) R_s = 0, R_{sh} = 100 and ∞; 5-11) R_s = 0, 1, 2, 3.5, 5, 10, and 20 Ω, respectively.

Figure 1.19 shows a typical current-voltage characteristic. The light characteristic was recorded under AMO solar simulator, and the dark characteristic was measured with an external bias in the forward (fourth quadrant) and reverse (second quadrant) direction. The portion of the light characteristic lying in the first quadrant, and its continuation in the fourth quadrant, constitute a straight line. The slope of this line is equal to the series resistance of the solar cell, the slope being measured in the region near U_{oc}: $R_s = \Delta U_f / \Delta I_f$.

The portion of the curve lying in the first quadrant and its continuation in the second quadrant also lie on the straight line. Its slope is equal to the shunt resistance of the cell, measured in the region near I_{sc}.

Since it is difficult to measure the slope of the straight line on the light characteristic near the point I_{sc}, the shunt resistance is usually determined from the slope of the dark characteristic (dashed line in the second quadrant):

$$R_{sh} = \Delta U_{rev} / \Delta I_{rev} .$$

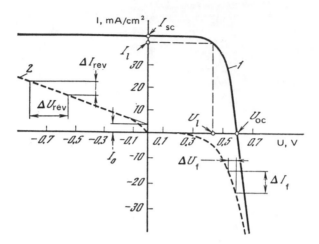

FIG. 1.19. Typical current-voltage characteristic of a modern silicon solar cell used in measurements under AM0 solar simulator: 1) light; 2) dark.

The reverse saturation current I_0 can also be found from the dark characteristic. It is given by the segment of the ordinate axis between the origin and the point of intersection with the continuation of the linear part of the reverse branch of the dark characteristic.

Since, however, the p–n junction of a solar cell is normally used in the forward direction (the effect of the incident radiation and the appearance of the associated excess nonequilibrium carriers on either side of the p–n junction are analogous to the application of a voltage in the forward direction), it is more correct to determine the saturation current and to calculate the coefficient A from the forward branch of the dark characteristic or from the light current-voltage characteristic (which is also called the *load characteristic* of the cell).

In the first of these methods, we can use the dark diode characteristic (1.15) in the form

$$\ln\,(I_D + I_0) = \ln I_0 + \frac{q}{AKT}\,U.$$

This equation is used only in the case of large currents ($I_D \gg \gg I_0$), and also for the recombinational mechanism of the reverse saturation current through the p–n junction [70], which requires the introduction of the coefficient A, as already noted in connection with (1.15). The segment corresponding to high currents and voltages (typical for the working load point of the solar cell) on the straight branch of the dark characteristic is used to construct the function $\ln I_D = f(U)$. The slope of this line is equal to q/AKT and the intercept on the ordinate axis is equal to $\ln I_0$.

There is one further method of determining I_0 and A under conditions approaching the working conditions of the solar cell. In this approach, the current-voltage characteristic is measured for a number (at least two) different incident-radiation intensities from the solar simulator.

Let us rewrite (1.14) and (1.15) so that they explicitly show the voltage drop across the series resistance and the terms representing recombination in the p–n junction:

$$I = I_0(\exp(q(U-IR_n)/AKT)-1)-I_{ph}. \qquad (1.18)$$

Under open-circuit conditions, $I = 0$, $U = U_{oc}$, and, when $R_s = 0$, the photocurrent I_{ph} may be assumed to be equal to I_{sc}. We then have

$$\ln(I_{sc}+I_0) = \ln I_0 + qU_{oc}/AKT.$$

For each new value of the radiation intensity from the solar simulator, which is determined with a reference solar cell having a linear relationship between the short-circuit current and the irradiance, a measurement is made of I_{sc} and U_{oc}, and a graph is plotted of $\ln I_{sc}$ as a function of U_{oc}. The slope of this line is equal to q/AKT and the intercept is $\ln I_0$.

The light characteristics can thus be used to determine A and I_0, and these values are typical for solar cells under operating conditions.

An improved and at the same time relatively simple way of determining R_s, R_{sh}, A, and I_0 was proposed in [71]. It relies on the measurement of the current-voltage characteristic of a solar cell for a single value of the incident intensity. This is then used to determine the area P_0 under the $U = f(I)$ curve. The slope of this curve at $I = 0$ is $\tan \alpha = AKT/qI_{sc} + R_s$. The next step is to measure the area P_1 and the graph of the cell power IU as a function of the current I. Subsequent calculations are based on the formulas given in [71].

Solar cell design

The above discussion of the basic processes occurring in the solar cell when it converts optical radiation into electric power clearly shows that the efficacy of these processes depends on the optical and electrophysical properties of the semiconducting material (reflection from the surface, photoionization quantum yield, diffusion length of minority carriers, position of the fundamental absorption band), as well as on the characteristics of the p—n junction (origin of reverse current, height of potential barrier, width of the space charge region), the so-called geometric factor (ratio of carrier diffusion length and depth of the p—n junction), and the dopant concentration on either side of the p—n junction. It is clear from (1.16), (1.17), and (1.18) that the shape of the current-voltage characteristic and the output power depend on the series resistance which, in turn, depends on the resistance, thickness, and dopant concentration in the semiconductor, as well as on the shape and position of the current contacts. The desire to reconcile these frequently conflicting requirements, and to find the optimum technological compromise, has led to the preferred use of the planar design for the solar cell (see Fig. 1.15a). With small modifications (introduction of drift fields and of an isotype barrier on the back contact; replacement of a solid back contact with a grid; texturing of the surface of the semiconductor and of its coating; and deposition of a reflecting layer on its rear surface), this design has been with us without any substantial change for many years, at any rate in the case of solar cells made from single-crystal silicon with a homogeneous p—n junction, and continues to dominate space and terrestrial applications.

Radiation-shielding, temperature-regulating, and antireflective coatings for solar cells are described in detail in [23]. The "working" surface of the silicon solar cell, which faces the incident optical radiation, is very thin and highly doped (impurity atom concentrations up to 10^{20}-10^{21} cm^{-3}), for example, with phosphorus atoms, so that it becomes an n-type region. The p-type base region of the semiconductor is usually lightly doped, for example, with boron (usually, while the crystal is being grown), up to impurity atom concentration of 10^{16}-10^{17} cm^{-3}. The outer surface of the solar cell is covered by a grid of current-taking strips occupying 5-7% of the total area [5,13,21], while a solid or grid contact is placed on the rear of the cell.

The minority carriers separated by the p—n junction field must enter the external circuit (load). In the upper, n-type, region of the semiconductor, which faces the incident light, the minority carriers move along the layer, whereas in the p-type base region (see Fig. 1.15a) they move across the layer. The diffusion length of the minority carriers in the highly doped n-type upper layer is usually 0.2-0.6 μm, while in the base layer it is 100-200 μm. These figures depend on the impurity concentration and the thermal treatment (number of cycles, rate of heating and cooling, maximum temperature) applied to the crystal while it is being grown, and to the solar cell during the fabrication process (for example, during the thermal diffusion of dopants and the deposition and strengthening of antireflective coatings).

The effect of thermal treatment on the properties of semiconducting materials and solar cells is investigated in [72,73], and the possibility of reducing this effect by gettering undesirable impurities from the silicon base region and of establishing strict control of thermal treatment is examined in [74]. These questions are discussed in [19] in relation to gallium arsenide solar cells.

It is important to note that the numerous thermal treatments applied to the semiconducting layers at different technological stages of solar-cell fabrication necessarily involve the entry of undesirable impurities and recombination centers which affect the optical and electrophysical parameters of the semiconducting material. This means that the parameters of the semiconductor are best determined at the end of the technological cycle. This is usually done by calculation from the output characteristics of solar cells, e.g., the current-voltage characteristic or the spectral sensitivity, or from certain other more specific curves, for example, the capacitance-voltage and current-irradiance characteristics (capacitance as a function of applied voltage and basic photoelectric parameters as functions of irradiance). These characteristics are usually measured in those cases where solar cells are used in automation or optoelectronics, so that rapidity of response and linearity at low and high irradiance levels are important [17,18].

The diffusion length in the doped layer is small, so that we have to use a shallow p—n junction (0.3-0.6 μm in modern mass-produced solar cells). To ensure that all the incident solar photons with $h\nu \geqslant \geqslant E_g$ are absorbed, the thickness of the base region must be not less than 200 μm. The resistance of the base region is low: current flows across a layer of relatively large cross section, toward the solid or grid base contact fused into the silicon at 750-800°C in an inert atmosphere. The first layer of the contact is often made of aluminum, which is a p-type impurity, in order to reduce the metal-silicon junction resistance (p-type). Aluminum is deposited by evaporation in a high vacuum, or in the form of aluminum-containing paste with an organic binder. The aluminum layer is then covered by a film of titanium, palladium, or silver (nickel is an alternative choice), and a layer of tin or lead solder [20,73].

The high layer resistance of the n-type top silicon layer which, as a rule, is in the range 50-100 $\Omega/\square$, is effectively reduced by placing on the outer surface a dense metalized grid of current contacts, made from the same material as the back contact (with the exception of the aluminum layer, which is unnecessary in the case of contact with the n-type layer). The configuration of the contact grid can be calculated from the formulas given in [13,75, 76]. Another problem that is encountered during the fabrication of the top current contact is that it is essential to produce a satisfactory (nonrectifying) contact that will not pierce the very thin doped layer during deposition and subsequent treatment. Experiment shows that the deposition of the metal layer over the entire outer surface, followed by the formation of the contact figure by etching, gives rise to the appearance of shorting microregions, which reduce R_{sh}, and to an increase in I_0 in both single-crystal and thin-film solar cells. This can be avoided by depositing the contact strips through a metal mask [77], or through windows in a layer of a polymeric photoresist or antireflective coatings, or directly through the antireflective coating [78]. At any rate, it is essential to ensure that the metal and the doped layer come together only at the intended point of contact.

For a layer resistance of between 50 and 100 $\Omega/\square$ on the outer surface of a 2 × 2 cm silicon solar cell, it is sufficient to produce one contact in the form of a strip, 0.5-1.0 mm wide, on one side of

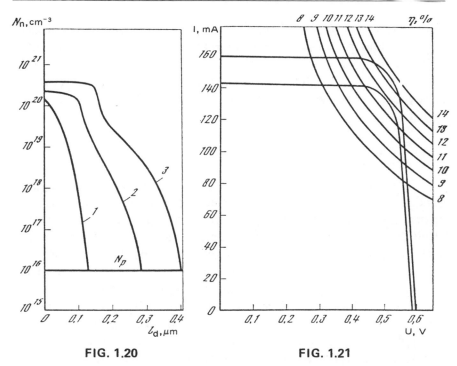

FIG. 1.20   **FIG. 1.21**

FIG. 1.20. Free-carrier concentration as a function of depth in the upper highly doped layers of a modern silicon solar cell for different positions of the p–n junction: 1-3) 0.12, 0.28, and 0.4 µm, respectively.

FIG. 1.21. Current-voltage characteristics of two modern silicon solar cells (2 X 2 cm), exposed to an AM0 solar simulator (irradiance 1360 W/m^2) and lines of equal efficiency.

the cell, with between 6 and 12 outgoing current-collecting contact strips, 0.05–0.1 mm wide, attached to the main strip. This reduces the contribution of the doped layer the total series resistance R_s of the cell down to 0.15–0.2 Ω. However, for very shallow p–n junctions (l = 0.15–0.4 µm), similar to those whose diffusion profiles (impurity concentration as function of depth) are shown in Fig. 1.20 [79], the layer resistance rises to 500 Ω/□, and the number of contact strips on a 2 X 2 cm solar cell must be increased to 60 (the necessary low resistance of a 15–20- µm contact strip is then achieved by subsequent electrochemical deposition of a silver layer of thickness up to 3–5 µm). When the contact figure on the surface of a planar solar cell is produced by high-precision technology in accordance with calculations, the current-voltage characteristics can be much improved [79] (they become almost square) and the cell efficiency η under the AMO solar simulator turns out to be 12-13.5% (Fig. 1.21).

Several new materials for contacts on doped layers have recently been proposed, for example, titanium nitrides, which have a negligible junction resistance on silicon.

Optimum semiconducting material and spectral sensitivity

The solar cell with a p—n junction in a homogeneous semiconductor is made from a homogeneous semiconductor whose basic optical and electric properties (including the band gap) are the same at all points within its volume. Structures and solar cells based upon them are referred to as *graded-gap* systems if the width of the band gap varies, e.g., decreases with distance into the crystal, owing to the continuous variation in the chemical composition of the material, and the p—n junction lies at a certain depth. The junction may be located on the boundary between two semiconducting layers of different band gap (it is then called a *heterojunction*), or it may be in one of the layers, usually in the lower layer with the smaller band gap. The upper layer of the wide-gap material is then merely an optical window that transmits the incident light onto the p—n junction. On the other hand, the boundary between the wide-gap and narrow-gap materials with similar lattice constants (as in the case of GaAlAs—CaAs and Cu_2S—CdZnS) has a low rate of carrier recombination. Since in solar cells with a p—n junction in a heterostructure, recombination on the upper boundary turns out to be sharply reduced, the carrier collection efficiency (especially in the short-wave part of the spectrum) is higher, and the efficiency of such cells can be very high [80].

During the early stages of the development of the homogeneous solar cell it was considered that it was desirable to use a semiconductor with a band gap equal to the photon energy corresponding to maximum intensity in the solar spectrum, i.e., approximately 2 eV [81]. It became clear subsequently that a reduction in the band gap leads to an increase in the number of photoactive solar photons and an increase in I_{sc}, but the photo-emf generated by such cells is then reduced because of the lower height of the potential barrier of the p—n junction. The dependence of the possible efficiency on

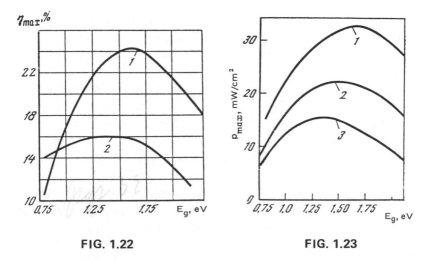

FIG. 1.22   FIG. 1.23

FIG. 1.22. Maximum efficiency of a solar cell under AM0 conditions as a function of the gap width of the semiconducting material: curve 1) the coefficient A in the expression for the reverse saturated current [see (1.16)] equal to unity; curve 2) A = 2.

FIG. 1.23. Maximum specific power generated by a solar cell as a function of the gap width of the semiconducting material under different conditions of absorption of solar radiation: 1) air mass m = 0 (AM0), thickness of precipitated water vapor in the atmosphere ω = 0, 2) m = l (AM1), ω = 2 cm (with selective absorption bands); 3) m = 3 (AM3), ω = 0.

the band gap of the semiconductor can be obtained only as a result of an analysis of the entire current-voltage characteristic of the solar cell and of the effect upon it of the spectrum of the incident radiation.

This calculation was first made in [82], using the spectra of solar radiation recorded on the Earth's surface [83]. Estimates of the optical and photoelectric losses were found to be very close to those for solar cells with optimum semiconducting materials. Subsequent calculations of maximum cell efficiency have led to a number of relationships (some of which are illustrated in Figs. 1.22 and 1.23 [82]).

Analysis of the results of these calculations has pointed up ways of making solar cells from semiconducting materials other than silicon. The most suitable semiconductors for maximum efficiency exceeding the efficiency of silicon solar cells are those with E_g in the range 1.1-1.6 eV (see Figs. 1.22 and 1.23).

For terrestrial solar radiation, the band gap of the optimum semiconductor can be smaller. A very important aspect from the standpoint of maximum efficiency of photoelectric energy conversion is the mechanism responsible for the reverse current through the p—n junction, which determines the coefficient A and the value of I_0. Improvements in the values of these p—n junction parameters may lead to a more substantial rise in solar-cell efficiency (see Fig. 1.22) as compared with the expansion of the spectral range of photoactive absorption of solar radiation by the semiconducting material.

In a solar cell with a p—n junction in a homogeneous semiconducting material, the junction collects and separates light-generated minority carriers produced on either side of the junction, i.e., both in the n- and p-type regions. This also occurs in most other more complicated versions of the solar cell with the exception, probably, of those cases where the charge carriers are separated in the metal-semiconductor barrier (Schottky barrier) and one of the regions is photoactive or completely absorbing for the entire solar radiation (to a considerable extent, this occurs in thin-film solar cells based on the copper sulfide-cadmium sulfide heterostructure in which by virtue of the high absorption coefficient of copper sulfide, practically the entire solar radiation is absorbed although the thickness of the copper sulfide layer is usually small, ranging from 0.05 to 0.2 μm).

It was shown above that, in the fundamental absorption band of the semiconductor, which defines the range of spectral sensitivity of a solar cell made from this material, the photoionization quantum yield is $\beta = 1$. Consequently, the effective quantum yield Q_{eff} of a solar cell and the carrier collection coefficient γ amount to practically the same thing [see formula (1.12)], so that both quantities will now be denoted in the same way, and will be represented by the collection coefficient Q.

The collection coefficient (the ratio of the number of excess carriers separated by the p—n junction to the number of electron-hole pairs produced by light) is the sum of the carrier collection coefficients of the p- and n-type regions on either side of the p—n junction:

$$Q_z=Q_n+Q_p=I_{sc}/(1-r)\,qN_l,$$

where I_{sc} is determined by the sum of the electron and hole currents from the p- and n-type regions, and the distribution N_l of solar-light photons with depth l in the semiconductor can be calculated from (1.3), assuming a wavelength dependence of the absorption coefficient $\alpha(\lambda)$, which is known for the given semiconductor.

The results of such calculations (performed by T. M. Golovner and G. A. Gukhman), using known $\alpha(\lambda)$ for silicon and gallium arsenide, are shown in Fig. 1.24.

For qualitative estimates of carrier collection from different portions of the solar cell or photodetector it is also useful to have compact formulas for Q and I_{sc} that are suitable for use on a computer [84], and also the following data on the penetration depth in silicon for optical radiation of different wavelength [17]:

λ, μm	0.45	0.5	0.56	0.6	0.65	0.70	0.75
$1/\alpha$, μm	0.4	0.89	1.61	2.12	3.06	4.33	6.14
λ, μm	0.80	0.85	0.90	0.95	1.0	1.05	1.1
$1/\alpha$, μm	8.9	14	24	63	208	2000	4000

Note. The last two values of $1/\alpha$ were calculated from Figs. 1.1 and 1.24.

The formulas for $I_{sc}(\lambda)$ and $Q(\lambda)$ derived in [13,80,81,84,85] are less convenient for computer calculations, but are easier to handle in calculations of the individual optical and electric parameters of semiconducting materials, especially when they can be simplified in the light of existing information. As a rule, these parameters undergo considerable changes in the course of the numerous thermal treatments applied during the prolonged process of manufacture of solar cells. The starting point for the derivation of these formulas is provided by the equations of continuity, written with and without taking the field into account. The equations contain terms describing the increase in minority carrier concentration in the semiconductor during diffusion from the ambient medium, which determine the amount of minority carriers lost by recombination and describe the generation of excess minority carriers by light, using a generation function consistent with (1.3) and reflecting the

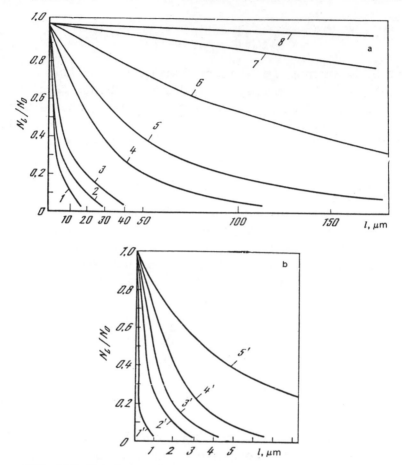

FIG. 1.24. Relative number of photons reaching a layer of depth l in silicon (a) and gallium arsenide (b) for different wavelengths: 1) 0.5 μm; 2) 0.7; 3) 0.8; 4) 0.9; 5) 0.95; 6) 1.0; 7) 1.05; 8) 1.1; 1') 0.4; 2') 0.5; 3') 0.7; 4') 0.8; 5') 0.9 μm.

influence of the electrostatic field and its gradient.

The component of I_{sc} due to the electron diffusion current through the p—n junction is given by the following expression for a p-type base (it is assumed that the distribution of impurities in the base of the solar cell is uniform and there is no drag field):

$$I_{sc\ b} = q\,\alpha\,L_n N_0 \exp\left(-\alpha l_d\right)/(1 + \alpha L_n) \qquad (1.19)$$

where in the case of an n-type base layer, L_n must be replaced with L_p.

Several researchers have discussed different ways of determining the individual parameters of a solar cell under certain simplifying numerical and experimental assumptions. For example, one method of estimating the diffusion length of minority carriers in the doped layer involves a comparison of calculated (for different l_d/L) and experimental distributions of the collection coefficient in the short-wave part of the spectrum in the case where the depth l_d of the p—n junction has been determined in advance [86]. The above data on the penetration of optical radiation in silicon at different wavelengths can then be used together with Fig. 1.24 to determine quite readily the wavelength of optical radiation that should be used in such experiments in order that the excess carriers are produced mostly in the upper doped layer of the cell. The depth of the p—n junction can be determined with adequate accuracy, using layer-by-layer anode oxidation and etching [see (1.3)] and (1) estimating the color of an oblique plane or cylindrical section (cut at a small angle to the surface of the cell, usually about 3°) [13], (2) finding the transmission or reflection coefficient of the p—n junction in the infrared part of the spectrum (see Fig. 1.10) with allowance for the depth of penetration of light, or (3) approximately calculating the depth from the surface layer resistance [38,50]. Similar methods of estimating the parameters of the doped layer are proposed in [84, 87]. By analyzing the individual (mostly short-wave) portions of the spectral dependence of the collection coefficient, it is possible to estimate the ratio S/D and, if the diffusion coefficient D is known, determine the rate of surface recombination S and the parameters l_d and L_p in the doped layer [88-90].

The most reliable way of determining the diffusion length of minority carriers in the base layer of a solar cell (p-type base layer) is based on (1.19) [91-93]. Since in the long-wave part of the spectrum (wavelength around 1 μm) we can neglect absorption in the doped layer in the case of measurements of I_{sc} and Q, these two quantities are determined by the base layer. For example, the spectral dependence of the collection coefficient is given by

$$Q(\lambda)=\alpha L_n \exp(-\alpha l_d)/(1+\alpha L). \qquad (1.20)$$

In modern solar cells, $l_d \cong$ 0.15-0.5 μm and for λ = 1 μm we have α_{Si} = 80 cm^{-1}, so that the term $\exp(-\alpha l_d)$ is very close to unity. Formula (1.20) is then simplified still further:

$$L_n = Q(\lambda)/\alpha(1-Q(\lambda)).$$

Having measured I_{sc} and the reflection coefficient at λ = 1 μm, and knowing N$_l$ (see Fig. 1.24) and q, we can readily determine Q at λ = 1 μm, and then L$_n$. To obtain a more accurate result, the necessary measurements are best performed at three closely spaced wavelengths (for example, 0.95, 1.0, and 1.05 μm) and the average diffusion length is calculated for the three measurements.

A more complicated situation arises when L in both the doped and base layers varies with depth as a result of increased efficiency due to the drag field deliberately introduced into the system, or the nonuniform distribution of radiation or thermal defects. The true diffusion length of the base layer region exposed to radiation can be determined if we know the effective diffusion length (the resultant for the damaged and undamaged portions of the base layer) and the diffusion length in the undamaged material [94,95]. The effect of nonuniform damage to the base layer on the spectral sensitivity is investigated in [96,97].

An interesting method was proposed in [98] for determining the depth of the p—n junction, or the diffusion length of minority carriers, from the position of the maximum of the spectral sensitivity curve, using the gallium arsenide cell as an example. A simple and convenient graphical method was put forward in [99]. It uses the carefully measured wavelength dependence of the collection coefficient and a number of simplifying assumptions (uniform internal electric field and constant carrier parameters) to estimate, with sufficient precision, solar-cell parameters such as the depth of the p—n junction, the rate of surface recombination, the carrier diffusion length, and the electric field for given geometry of the solar cell and given electrophysical properties.

The diffusion length of minority carriers in the base region, deduced from the measured optical and photoelectric parameters of solar cells, can be usefully compared with values determined by exposing solar cells to gamma rays [100] and electrons [101].

Method of determining the spectral sensitivity
and the collection coefficient of solar cells

The spectral sensitivity of a solar cell is the wavelength dependence of its short-circuit current per unit energy of the incident optical radiation.

Routine measurements of the spectral sensitivity are usually made with the ZMR-3 mirror monochromator with glass optics (Fig. 1.25). The source of light is the 36-W hot-filament lamp SI-6-40. An image of the filament is projected by an elliptic reflector onto the entrance slit of the instrument. The light intensity is held constant by holding constant the filament supply current (5.8 A), monitored with an ammeter.

The SI-6-40 lamp is suitable for measurements in the visible and infrared (up to the glass transmission limit). Measurements in the ultraviolet (0.4-0.3 μm) are made by using a hot-filament lamp and a uviol window. Its color temperature is 3200 K. Finally, at wavelengths below 0.3 μm, it is best to use a hydrogen lamp [102], which produces a continuous spectrum and is highly stable as compared with other gas-discharge sources.

The working slit of the ZMR-3 monochromator was varied between 1 mm in the range 0.4-0.5 μm and 0.25 mm in the range > 0.9 μm, so that the spectral slit width remained constant in the range 0.01-0.015 μm. The wavelength can be measured in the ZMR-3 monochromator by rotating the mirror 4 (Fig. 1.26) through a small angle, which allows radiation of any particular chosen wavelength to pass through the prism with minimum deviation from its incident direction.

Scattered light is removed in spectral sensitivity measurements (see Fig. 1.25) by using the SZS-8, SZS-14, or SZS-12 light filters in the respective spectral ranges. Two lenses are placed in front of the exit slit of the monochromator and are used to distribute the light flux over the entire surface, or to concentrate it on part of the solar cell. At a certain distance from the second lens, this light flux is projected onto a strip of 1.5 $\times$ 9 mm, which covers completely the receiving plate of a thermocouple. The latter is then replaced with the solar cell, so that the entire light flux intercepted by the thermocouple is, in turn, intercepted by the surface of the solar cell (see Fig. 1.25). If it turns out that, when the illuminated area is deter-

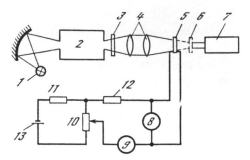

FIG. 1.25. Schematic diagram of apparatus used to measure the spectral sensitivity of solar cells: 1) source of light; 2) monochromator; 3) demountable optical filters; 4) lenses; 5, 6) working and reference solar cells, respectively (thermocouple in the case of calibration); 7) potentiometer; 8) voltmeter; 9) ammeter; 10-12) resistors R_1-R_3, respectively; 13) compensating voltage (E = 1.5 V).

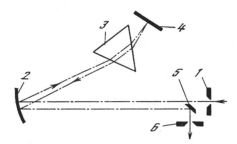

FIG. 1.26. Optical arrangement of the mirror monochromator ZMR-3: 1) entrance slit; 2) mirror objective; 3) prism; 4) rotating flat mirror (Littrow mirror); 5) flat mirror; 6) exit slit.

mined with the aid of a traveling microscope, a proportion of the incident flux falls on contact strips on the solar cell, the illuminated contact area can, of course, be subtracted from the total illuminated area.

The flux density of monochromatic radiation can be measured, for example, with the Kozyrev compensated vacuum thermocouple (type RTE, receiving plates of 2 × 12 mm). The thermocouple output is fed into a low-resistance potentiometer. The sensitivity of one of the thermocouples used in these experiments was 0.47 V/W and was periodically checked against standard lamps calibrated at stan-

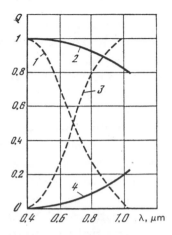

FIG. 1.27. Contributions of the doped (1, 2) and base (3, 4) regions to the resultant collection coefficient of solar cells: 1, 3) silicon (n-type on p-type); 2, 4) gallium arsenide (p-type on n-type).

dardizing laboratories at Leningrad and Moscow, where color temperature and flux density measurements were performed.

Nonselective radiation detectors that can also be used include those incorporating metal thermocouples [103], thin-film thermocouples [104], and semiconductor thermopiles [105]. These detectors can be calibrated by a number of independent methods, including the use of a standard lamp, or a substitution coil carrying the necessary current [105], or by using a black body maintained at a known temperature.

In these measurements, the entire spectrum produced by the monochromator between 0.4 and 1.16 μm is calibrated with a thermocouple which is then replaced by the solar cell whose short-circuit current is measured with the balanced circuit of Fig. 1.25. This procedure of solar-cell substitution after each wavelength measurement introduces significant positioning uncertainties.

The null indicator was the M195/2 galvanometer and the measuring device was the M95 microammeter with the necessary shunt. The following resistors were used: $R_2 \cong 100\ k\Omega$, $R_3 \cong 3\ k\Omega$, and R_1 up to 10 kΩ. This arrangement eliminates the effect of the microammeter resistance on the measured current. The photocurrent could be determined to within ±10% in the wavelength range 0.45-

0.5 μm and to within ±5% for $\lambda > 0.5$ μm. The uncertainty may be as high as 30% in the range 0.4-0.45 μm for solar cells with low sensitivity in this spectral range.

The energy leaving the monochromator was measured over the entire spectral interval in the range 0.002-0.02 mW (corresponding to a photon flux range $1.5 \cdot 10^{12} - 1 \cdot 10^{14}$ sec^{-1}).

It is important to note that, because the current-irradiance characteristic of many solar cells is nonlinear between the region of low irradiances produced by monochromatic light and the high irradiances due to AM0 solar radiation and, on clear days, on the Earth, precision measurements of the spectral sensitivity (e.g., of reference cells) are performed with special precision equipment.

The absolute spectral sensitivity is calculated after measuring $I_{sc}(\lambda)$ in the form of the ratio $I_{sc}(\lambda)/E(\lambda)$. To determine the collection coefficient (I_{sc} per absorbed photon), we must also determine the reflection coefficient of the solar-cell surface in the same spectral range.

Experiments and calculations have shown that the contribution of the top doped silicon layer to the resultant collection coefficient begins to decrease as the depth of the p—n junction is reduced and the rate of surface recombination increases [63,84].

This tendency is also observed for gallium arsenide solar cells. The homogeneous p—n junction in this material is usually produced by careful thermal diffusion of zinc (p-type impurity) into the original n-type gallium arsenide. However, the much higher absorption coefficient of gallium arsenide (as compared with silicon) and its rapid variation with wavelength (see Fig. 1.1) ensure that practically the entire photoactive radiation is absorbed in the top p-type layer and is collected from it by the lower lying p—n junction in gallium arsenide.

Comparative calculations have also been performed for silicon and gallium arsenide solar cells with similar layer thickness and the same rate of surface recombination. These calculations have shown that the contribution of the base layer to the resultant collection coefficient becomes appreciable only in the long-wave region of the spectral sensitivity of gallium arsenide solar cells (Fig. 1.27, curve 4).

The spectral sensitivity of the solar cells investigated was used in accordance with the method described in [86] to calculate the lifetime and diffusion length of minority carriers on both sides of

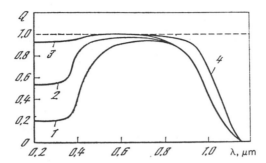

FIG. 1.28. Collection coefficient of silicon solar cells with different combinations of electrophysical parameters as a function of wavelength: 1-4) order number in Table 1.2.

the p—n junction [106]. It was found that the additional reason for the slight effect of the base region on the overall collection coefficient of gallium arsenide cells was the very small diffusion length of minority carriers in this material, which did not exceed 1.5 μm.

Limiting values of collection coefficient and spectral sensitivity

Infinite lifetime and infinite diffusion length of minority carriers in both layers of a solar cell and zero rate of surface recombination (as well as at R = 0) would correspond to a collection coefficient equal to unity across the entire photosensitivity range, and the spectral sensitivity curve would have a sharp peak at photon energy hν, equal to the gap width E_g of the semiconductor used to fabricate the cell. Thereafter, the spectral sensitivity would decrease linearly with increasing frequency (i.e., reducing wavelength) of the incident optical radiation [21].

When there is no surface or volume recombination, all carriers produced in the semiconductor by radiation of wavelength λ must be collected and separated by the p—n junction:

$$I_{sc}(\lambda) = qN_0(\lambda) = qE(\lambda)/h\nu.$$

Hence, it is clear that the spectral sensitivity is then a linear function of wavelength:

TABLE 1.1
Limiting Values of Spectral Sensitivity of a Planar Solar Cell
with a p—n Junction in a Homogeneous Semiconducting Material

λ, μm	I_{sc}/E, μA/mW		λ, μm	I_{sc}/E, μA/mW	
	$R_\lambda = 0$	$R_\lambda = R_{Si}$		$R_\lambda = 0$	$R_\lambda = R_{Si}$
0.3	240	96	0.80	640	435
0.4	320	170	0.85	680	460
0.45	360	250	0.90	720	500
0.55	440	280	0.95	760	525
0.60	480	315	1.0	800	548
0.65	520	344	1.1	880	607
0.70	560	370			

$$I_{sc}(\lambda)/E(\lambda) = q/h\nu = 0.8 \cdot 10^3 \, \lambda. \qquad (1.21)$$

When the reflection coefficient R of the surface of the solar cell is not zero, its wavelength dependence must be taken into account, and the spectral sensitivity is given by a nonlinear function of wavelength:

$$I_{sc}/E = 0.8 \cdot 10^3 \, (1 - R_\lambda)\lambda.$$

The limiting values of the spectral sensitivity of a semiconductor solar cell of planar design under the idealized conditions indicated above (zero rate of surface recombination, infinite lifetime, and infinite diffusion length of minority carriers) were calculated from (1.21) (Table 1.1) for two values of the reflection coefficient, namely, R = 0 and R equal to the reflection coefficient R_{Si} of polished, uncoated silicon (Fig. 1.11, curve 4).

The long-wave edge of the spectral-sensitivity curve of solar cells is governed only by the position of the fundamental absorption edge (or, as it used to be called, the red limit of the photoelectric effect), determined by the gap width of the semiconductor and the nature of optical band-band transitions. The left-hand edge of the sensitivity curve of a planar solar cell is largely determined by the rate of recombination on the surface collecting the incident radiation.

TABLE 1.2

Parameters of Silicon Solar Cells whose Wavelength Dependence
of the Collection Coefficient is Shown in Fig. 1.28

No.	l_d, μm	S, cm·sec^{-1}	τ_n, μsec	Q_Σ
1	0.4	10^5	3	0.71
2	0.1	10^5	3	0.79
3	0.2	10^2	3	0.83
4	0.2	10^2	12	0.88

Note: The thickness of each cell is 0.3 mm.

This situation is clearly illustrated by the calculated wavelength
dependence of the collection coefficient of solar cells, shown in
Fig. 1.28 [107] for different combinations of physical parameters
(Table 1.2). Analysis of these data leads to a number of conclusions
with regard to possible ways of improving solar cell technology.
These are:
   (a) the sensitivity can be extended toward longer wavelengths
       by increasing the lifetime of minority carriers in the base
       layer, for example, by using purer and higher resistivity
       original semiconducting material and maintaining its prop-
       erties during the fabrication of the solar cells;
   (b) silicon-based solar cells can be fabricated with very high
       sensitivity at short and ultraviolet wavelengths, down to
       0.2 μm (see Fig. 1.28), by sharply reducing the rate of
       surface recombination and the depth of the p–n junction.
   Thus, studies of the spectral sensitivity and collection coefficient of solar cells are exceedingly useful in attempts to (1) improve
further the properties of solar cells, (2) increase their efficiency, and
(3) consequently extend their range of application. Careful studies
of these parameters are also essential for achieving the necessary
precision of efficiency measurements, and for improving metrological data obtained by comparing reference and working solar cells. A
clear understanding of the reasons for parameter deviations is essential if these discrepancies and the associated experimental errors are
to be minimized.

2

Solar Cells with Improved Optical and Photoelectric Characteristics

The optical and photoelectric characteristics of solar cells can be improved in a variety of ways, e.g., by creating drift fields in the doped and base layers (for example, by suitably varying the impurity distribution or the gap gradient within the cell), by replacing homogeneous with heterogeneous structures, and by replacing the p—n junction with a metal—semiconductor barrier (Schottky barrier) or a metal—dielectric (usually an oxide layer)—semiconductor structure to collect excess charges. The thin metal layer is often replaced in these systems with the more transparent films of doped wide-gap semiconductors based, for example, on SnO_2, Cd_2SnO_4, or In_2O_3 —SnO_2 mixtures, frequently referred to as ITO films.

Thin-film solar cells have a number of interesting properties. Such films have to be developed, first of all, in order to reduce the cost of solar cells by reducing the amount of semiconducting material necessary for their fabrication. Thin-film cells are usually made of semiconducting materials with direct optical transitions and, as already noted, they have enhanced sensitivity at shorter wavelenghts, so that they can be used effectively as UV detectors with small linear dimensions.

The spectral range of solar cells can be substantially extended by developing cascade, two-sided, and multijunction solar cells. It is only since the development of the solar cell transparent in the long-wave region of the spectrum (beyond the fundamental absorption band [108-111]) that it has been possible to pass from theoretical models [63] to the experimental development of cascade cells. These are now capable of efficiencies of between 28 and 35% under terrestrial conditions [112,113].

Very high efficiencies (between 17 and 28%) have also been obtained by using fundamentally different physical principles to enhance the efficiency of conversion of solar radiation into electric power. They include:

(a) preliminary decomposition of the solar spectrum into two or more spectral intervals with the aid of multilayer beam splitters (dichroic mirrors), followed by the conversion of each spectral interval by its own high-efficiency solar cell [114];

(b) the use of reradiating heterostructures (with nearly 100% internal quantum yield), which compress the wide incident spectrum so that it can be subsequently efficiently transformed into electric power by a homogeneous semiconducting material, for example, gallium arsenide [115];

(c) the use of a structure consisting of a p—n junction in a homogeneous material, equipped on its outer surface with an optical window made from a wide-gap semiconductor and having depth-dependent chemical composition, so that the gap width of the window decreases with decreasing distance from the homogeneous material [116,117].

We shall now examine the characteristics of various practical photoelectric systems. We are interested, in the first instance, in the optical and photoelectric properties of solar cells with improved parameters and the physical processes responsible for these properties, as well as methods of modifying the sensitivity of solar cells in different parts of the spectrum.

2.1 High-efficiency silicon solar cells with a drift field in the doped region

The effect of internal electric fields on the collection efficiency and the efficiency of semiconductor solar cells has been investigated quite extensively (see, for example, [63,80,84,107,118,119]). Most of the initial work was concerned with the effect of a uniform field and constant mobility and carrier lifetime, which were independent of carrier concentration. The solar-cell models were then extended to built-in fields, and this led to the study of nonuniform electric fields and diffusion parameters that were functions of position [120]. However, these studies were essentially theoretical in character and the assumed carrier distributions were difficult to reproduce. Modern solar cells have internal electric fields that are random

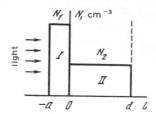

FIG. 2.1. Two-layer model of the doped region in a
solar cell.   The dashed line represents the p–n junction at a depth $l_d = a + d$.

in character and are a consequence of the fabrication technology
employed. The problem that had to be faced was therefore to find
the impurity distributions that would substantially enhance the
efficiency of collection of carriers from the doped layer and, at the
same time, could be produced by well-established technology.

Two-layer model of the doped top layer

We shall now consider the possibility of a solar cell in which the
doped layer consists of two regions with different impurity concentration (Fig. 2.1). A jump in the potential, $U_E = \ln(N_1/N_2)KT/q$,
occurs across the separation boundary between the two regions and
the concentrations obey the inequality $N_1 > N_2$, so that the electric
field across the separation boundary between regions I and II points
toward the p–n junction. The first step will be to optimize the
parameters of the doped layer with respect to the photocurrent and
power by taking into account the series resistance [121,122].
     The mobility and the carrier diffusion length as functions of
impurity concentration can be taken in the form

$$\mu_n, \ \mu_p \sim N^{-\gamma}, \ L_p \sim N^{-\beta},$$

which agrees with experimental data for $\beta = \gamma = 1/2$ to a sufficient
degree of precision. The final expression for the coefficient describing carrier collection from the doped layer when the rate of surface
recombination is $s \to \infty$ is as follows:

$$Q_d = \frac{\alpha \exp(-\alpha a)}{k_2 \exp(-2d/L_{p2}) - k_1} \times$$

$$\times \left(\frac{k_2 \exp(-d/L_{p2})(1 - \exp(-(\alpha + 1/L_{p2})d))}{1/L_{p2} + \alpha} + \right. \tag{2.1}$$

$$\left. + \frac{k_1(\exp(-\alpha d) - \exp(-d/L_{p2}))}{\alpha - 1/L_{p2}} - 2\exp(-d/L_{p2})k_3 \right),$$

where

$$k_{1,2} = (1 + \exp(-2a/L_{p1}))N_2/N_1 \pm (1 - \exp(-2a/L_{p1}))$$

$$k_3 = (\exp(-2a/L_{p1}) - \exp(\alpha a)\exp(-a/L_{p1}))/(\alpha + 1/L_{p1}) -$$

$$- (1 + \exp(\alpha a)\exp(-a/L_{p1}))/(\alpha - 1/L_{p1}),$$

L_{p1} and L_{p2} are the diffusion lengths of minority carriers in regions I and II, respectively, and a and d are the widths of these regions (see Fig. 2.1).

The integrated photocurrent drawn from the doped region when the cell is exposed to a perfect black body held at the solar temperature ($T_S = 6000$ K) is given by

$$j_p = \frac{15\hbar^3 q}{\pi^4 (KT_S)^4} E \int_{\omega_0}^{\infty} \frac{\omega^2 \exp(-\hbar\omega/KT_S)}{1 - \exp(-\hbar\omega/KT_S)} Q_d(\omega) \, d\omega, \tag{2.2}$$

where $\hbar = h/2\pi$ is Planck's constant, $\hbar\omega_0$ is the gap width (for silicon, $\hbar\omega_0 = 1.12$ eV), ω_0 represents the red limit of the photoelectric effect, $Q_d(\omega)$ is given by (2.1), and the solar constant is $E = 1360$ W/m^2.

In deriving (2.2), we used an approach similar to that employed in [123], where the Bose-Einstein equilibrium function was used to calculate the number of photons of energy $\hbar\omega$, emitted by the Sun as a perfectly black body into the solid angle subtended by the Earth. When the relationships given in [123] were employed, it was more convenient to use the angular frequency ω instead of the ordinary frequency ν, and to represent the gap width E_g by $\hbar\omega_0$.

The integral in (2.2) was evaluated numerically to a high precision [80].

The photocurrent from the doped layer was obtained as a function of the thickness of the region with enhanced carrier concentration for different values of a and d (Fig. 2.1). The dopant concentration in the p—n junction was $N_2 = 10^{17}$-10^{18} cm^{-3}. The carrrier concentration N_1 on the surface was assigned a number of values in the range 10^{18}-10^{21} cm^{-3}, where the maximum value $N_1 = 10^{21}$ cm^{-3} corresponded to the solubility limit of phosphorus in silicon.

Our calculations showed that the maximum photocurrent from the doped layer was obtained within the above limits for $a = 0.05$ μm and $N_1/N_2 = 100$.

However, subsequent analysis revealed that, for fixed N_1 and N_2, the useful power can be a maximum for $a > 0.05$ μm. The point was that, to produce a high photo-emf in real solar cells, the necessary carrier concentration N_2 in the p—n junction had to be 10^{17}-10^{18} cm^{-3}. For such concentrations, the spreading resistance of the thin (1.0 μm) doped layer was found to be relatively high, but could be reduced (for the same shape of contact on the working surface) by expanding region I with a higher impurity concentration.

Figure 2.2 shows the power produced by solar cells with anti-reflective coating and optimized doped-layer parameters described by the above model. The figure also shows the results for an ordinary solar cell, in which the doped layer has a uniform electric field (the concentration falls from $5 \cdot 10^{20}$ cm^{-3} on the surface to 10^{17} cm^{-3} at the p—n junction, the impurity being an exponential function of depth).

The impurity concentration of 10^{17} cm^{-3} corresponds to minority-carrier diffusion lengths $L_p = 1$ μm and majority-carrier mobility $\mu_n = 600$ cm^2/V·sec. The contact grid has cells of h = l_c = = 0.5 cm. The p-type base is assumed to be infinitely thick, with minority-carrier diffusion length $L_n = 100$ μm.

The power was calculated from

$$P = I_{ph} \ln (I_{ph}/I_0 + 1) AKT/q - I_{ph}^2 R_s, \tag{2.3}$$

where A = 2, $I_0 = 10^{-7}$ A,

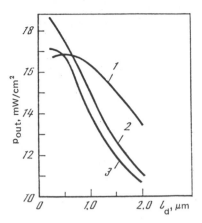

FIG. 2.2. Calculated output power as a function of the depth l_d of the p—n junction in a solar cell in which the electric field is: power optimized for $N_1 = 10^{19}$, $N_2 = 10^{17}$ cm^{-3} (1); power optimized for $N_1 = 10^{20}$, $N_2 = 10^{18}$ cm^{-3} (2); and uniform (3).

$$R_s = C/q\,(aN_1\mu_{n1} + dN_2\mu_{n2}),\tag{2.4}$$

μ_{n1} and μ_{n2} are the mobilities of majority carriers in regions I and II, respectively (see Fig. 2.1), and the coefficient C is given by [124]:

$$C = 64hl_c/\pi^6\,(l_c^2 + h^2).\tag{2.5}$$

The data shown in Fig. 2.2 lead us to the following conclusions with regard to the efficiency of solar cells with optimized two-layer doped top region (curves 1, 2) as compared with ordinary solar cells with a uniform electric field (curve 3).

When the impurity concentration in the doped layer is $N_1 = 10^{19}$ and $N_2 = 10^{17}$ cm^{-3} (curve 1), the power produced by the solar cell based on the above model exceeds the power delivered by an ordinary cell for $l_d > 0.6$ μm. This layer is particularly convenient in the case of a deep p—n junction. Actually, when $l_d = 0.7$ μm, the increase in power amounts to 5% and the corresponding figures for $l_d = 1.0$, 1.5, and 2.0 μm are 17%, 27%, and 28%, respectively. Doped layers with composite impurity distributions can be used to achieve higher useful power for greater depths of the p—n junction as compared with the exponential impurity distribution. For example, $P = 16$ mW/cm^2 corresponds to $l \cong 0.7$ μm (curve 3) and $l_d \cong 1.2$ μm (curve 1).

When the impurity concentration in the doped layer is raised to $N_1 = 10^{20}$, $N_2 = 10^{18}$ cm^{-3} (curve 2), the increase in power as compared with the case of the uniform field is 4-7% for all values of l_d. A somewhat greater increase in power (up to 10%) is observed for $l_d < 0.5$ μm. Thus, if solar cells can be made with reliable contacts for p—n junction depths less than 0.5 μm, better results can be achieved by producing the doped layer with a stepped distribution of high impurity concentration (to reduce the series resistance).

Fabrication of solar cells with a two-layer doped region

The above theoretical results were examined experimentally using a doped layer in which the stepped impurity distribution was established by thermodiffusion technology, widely used in the fabrication of silicon solar cells [2,5,13,79]. Diffusion was produced by the box method [124].

Calculations and experiments have shown that the porous oxide film initially produced on the surface of silicon by anode oxidation can be exploited even in the case of a single diffusion process to produce the two-layer doped region [125]. Part of the diffusant, for example, phosphorus, passes through the pores and forms a region of low-impurity concentration in the p—n junction. The impurity current retained by the oxide layer produces a thin layer with enhanced impurity concentration on the surface. By varying the porosity of the film, and controlling the diffusion time and temperature, it is possible to control quite smoothly and accurately the impurity distribution in the doped region.

A p—n junction in which the depth of the doped layer is 0.9-1.3 μm can be produced by optimized single thermodiffusion through an oxide film of the necessary porosity, which is first deposited on the system. The impurity distribution can thus be made to take the form of the two regions of high and low concentration (curve 1, Fig. 2.3).

Another possibility is double doping. This was achieved by using selected silicon disks with a 3-μm doped layer, produced by thermodiffusion, in which the impurity distribution was as shown by curve 3 in Fig. 2.3.

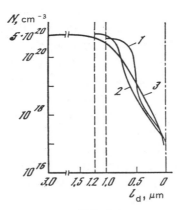

FIG. 2.3

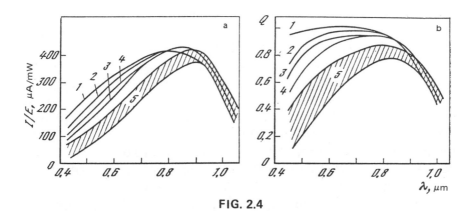

FIG. 2.4

FIG. 2.3. Measured concentration of phosphorus in silicon as a function of depth between the surface (dashed lines) and the p—n junction (dot-dash line). The layer is produced by thermodiffusion under different conditions: 1) single thermodiffusion through an impeding oxide layer of a particular porosity (l_d = 1.0 μm); 2) double thermodiffusion through the oxide layer on the surface (l_d = 1.2 μm); 3) thermodiffusion without preliminary oxidation of the surface (l_d = 1.2 μm after chemical etching of the doped layer).

FIG. 2.4. Spectral sensitivity (a) and collection coefficient (b) for uncoated solar cells fabricated by different methods: 1-3) as in Fig. 2.3; 4) shallow thermodiffusion in a gas flow (exponential distribution of impurities, l_d = 0.6 μm); 5) deep single thermodiffusion (without chemical etching after diffusion, l_d = 3.0 μm).

The diffused layer was etched out to a depth of 0.5-0.6 μm and this was followed by secondary doping by single thermodiffusion. The resulting p—n junctions were at a depth of 1.0-1.2 μm from the surface, and it was found that the impurity concentration changed by two orders of magnitude over the depth range 0.3-0.7 μm (curve 2, Fig. 2.3). The impurity concentration profile was determined from conductivity measurements, using the four-probe method and layer-by-layer anode etching. The depth of the p—n junction was determined by the grooving method.

Current contacts were deposited on the silicon wafers in the usual way [13,16,77,79], and the characteristics of the resulting photocells were investigated.

The experimental solar cells have enhanced spectral sensitivity in the short-wave part of the spectrum (curves 1 and 2, Fig. 2.4), which depends on the efficiency of collection of carriers from the doped region. For example, for λ = 0.5 μm, the experimental solar cells with stepped impurity distribution in the doped layer have I/E = 220-250 μA/mW whereas for cells with doped-layer thickness of the order of 3 and 1.2 μm (the corresponding impurity distributions are described by the horizontal and sloping parts of curve 3 in Fig. 2.3) the spectral sensitivity lies in the range 50-125 μA/mW (region 5, Fig. 2.4) and 170-180 μA/mW (curve 3, Fig. 2.4). Even for elements with a very shallow p—n junction (0.6 μm) and exponential impurity distribution (curve 4, Fig. 2.4), the sensitivity at λ = 0.5 μm does not exceed 200 μA/mW. Solar cells with p—n junction depth l_d = 0.6 μm (curve 4), produced by low-temperature diffusion, have a near-exponential impurity distribution in the doped layer with a concentration drop from $5 \cdot 10^{20}$ on the surface to 10^{16} cm^{-3} at the p—n junction. Comparison of curves 1-4 will show that the enhanced sensitivity of the experimental solar cells in the short-wave part of the spectrum (curves 1 and 2) can be explained by the dominating effect (as compared with the deterioration in the minority-carrier diffusion parameters in the region of enhanced concentration) of the built-in drift field of complex configuration.

The current-voltage characteristics of solar cells with the two-layer structure of the doped region are also much better than those of ordinary cells. The load current drawn per unit useful area of the solar cell when the p—n junction is at a depth of 1.0-1.2 μm is greater by 9-17% as compared with a cell using an exponential impurity distribution in the doped layer, and this can be regarded as a sufficient confirmation of the calculated results (see Fig. 2.2).

The proposed stepped distribution of impurities is thus seen to result in a considerable improvement in the current-voltage and spectral characteristics of solar cells, even when the p—n junction is relatively shallow ($l_d \cong 1.2~\mu m$), so that one can increase not only the efficiency of the cells but use simple, cheap, and reliable contacts produced by chemical deposition of nickel [13,16]. The problem of producing reliable ohmic contacts and of automating their deposition (as well as making them cheaper) is one of the most complex problems in modern technology of solar-cell fabrication.

2.2   Silicon solar cells with a passivating surface film

The short-wave spectral sensitivity of silicon solar cells can be substantially increased by using a passivating film, for example, a film of silicon dioxide or silicon nitride. The film contains a built-in electric charge and, together with the thin doped layer of silicon (as in solar cells with a drift field in the doped region; see Section 2.1) forms a two-layer structure of the form n^+—n or p^+—p, which brings up the electrostatic drift field closer to the surface, reduces the effective rate of surface recombination, and improves the collection of excess carriers produced by the short-wave radiation absorbed near the surface of the solar cell.

This type of n^+—n structure was produced on single-crystal silicon with resistivity of 0.1-0.3 $\Omega \cdot cm$ by bombarding it with 10-keV phosphorus ions, using an ion-beam density of between $2.5 \cdot 10^{12}$ and $2.5 \cdot 10^{15}$ cm^{-2} [126]. The implanted phosphorus was activated by following the ion implantation with thermal annealing for 30 minutes at 850°C in an atmosphere of water vapor and oxygen. This gave rise to the appearance on the surface of a 2000-Å silicon dioxide film, doped with phosphorus and boron drawn from the substrate. The integrity of the silicon dioxide film was improved by subjecting it to further thermal annealing in dry oxygen for one hour at 700°C (this resulted in an increase in the refractive index of the film to 1.48). To restore the lifetime of minority carriers in the base (substrate), the high-temperature annealing was followed by heating of the specimens to 550°C and holding them at this temperature for two hours. Slow etching was then used to produce a 1000-

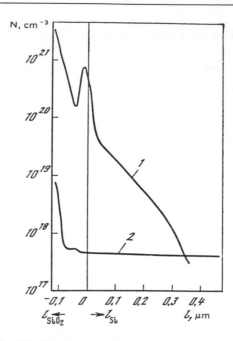

FIG. 2.5. Distribution of impurity concentration with depth in the doped surface film of silicon dioxide (l_{SiO_2}) and the top doped silicon layer (l_{Si}) in solar cells with a p—n junction produced by phosphorus ion implantation, followed by thermal annealing: 1) phosphorus; 2) boron.

Å silicon dioxide film, this being the optimum thickness for an anti-reflective coating. Finally, photolithography was employed to produce windows for contact strips with the usual three-layer composition (titanium—palladium—silver).

The distribution of the phosphorus and boron impurities in the doped SiO_2 film and in the top layer of silicon (Fig. 2.5) was determined by secondary-ion spectroscopy. The shallow p—n junction was found to be located at a depth of 0.35 μm.

As in the case of controlled diffusion through an oxide film, the impurity distribution was found to have the two-step profile (with a small jump in impurity concentration across the SiO_2—Si separation boundary; Fig. 2.5). Consequently, here again, we have an electrostatic drift field of enhanced strength, which is confirmed by the high collection coefficient of these solar cells in the short-wave

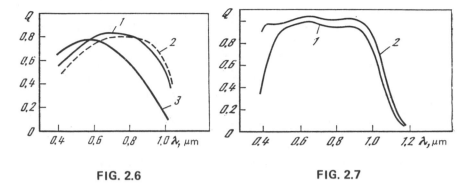

FIG. 2.6   FIG. 2.7

FIG. 2.6. Wavelength dependence of the collection coefficient of silicon solar cells produced by phosphorus ion implantation in silicon wafers [1, 2) zone melting without crucible; 3) Czochralski method] with different resistivity: 1) 0.3 $\Omega \cdot$cm; 2, 3) 0.1 $\Omega \cdot$cm.

FIG. 2.7. Spectral dependence of the collection coefficient of silicon solar cells before (1) and after (2) the deposition of the thin passivating film of silicon dioxide.

region of the spectrum (Fig. 2.6). The long-wave sensitivity of solar cells produced from high-resistivity silicon fabricated by zone melting (Fig. 2.6, curves 1 and 2) was quite high. On the other hand, in the case of low-resistivity silicon grown by the Czochralski method, the above thermal treatment was not optimal, so that the lifetime and diffusion length of minority carriers in the base of the final cell were low and the long-wave sensitivity was also low (curve 3, Fig. 2.6).

Solar cells produced by the above method have the following parameters under illumination by AMO radiation: η = 12.3-14.5%, I_{sc} = 34.5 mA/cm^2, U_{oc} = 0.645 V.

The unusually high value of U_{oc} can be explained by the high barrier across the p—n junction due to not only the low resistivity of the substrate for the base layer, but also the influence of the doped surface film. This was confirmed by direct experiment, whereby the film was removed by etching and U_{oc} was found to fall to the usual level (less than 0.6 V).

The influence of the passivating surface film on the collection coefficient at short wavelengths (Fig. 2.7) and on U_{oc} of solar cells was demonstrated experimentally in [127]: both these parameters were found to increase when the passivating film was introduced. The depth of the p—n junction under the passivating film was 0.3 μm for a layer resistance of 60 $\Omega/\square$ (the p—n junction was produced by diffusing boron into the phosphorus-doped substrate of n-type silicon, 300 μm thick). An antireflective coating of silicon nitride (SiN$_x$) was deposited on top of the passivating SiO$_2$ film. For an irradiance of 25 solar constants, the efficiency of the solar cells with the p$^+$—p structure on the illuminated surface was 18%. Under an irradiance of one solar constant, I$_{sc}$ was found to be 33 mA/cm^2 and U$_{oc}$ was 0.62 V.

2.3   Silicon solar cells with a drift field in the base and an isotype barrier on the back contact

When the base layer of a solar cell, for example, a p-type layer, is doped nonuniformly and the acceptor concentration in the p—n junction is lower than that within the body of the layer, an electric field is found to appear and assists in the collection of excess carriers produced by light in the base layer (both the diffusion and drift collection mechanisms are then found to be operative). It has frequently been noted [12,80,84] that an impurity gradient is a precondition for a drift field. On the one hand, this reduces the open-circuit voltage due to the increase in the reverse saturation current when the potential barrier is reduced (by reducing the dopant concentration in the base of the p—n junction) and, on the other hand, leads to a considerable deterioration in the diffusion length and lifetime of minority carriers (when the dopant concentration is increased in base layers that are distant from the p—n junction). Both phenomena work against the improvement in the collection coefficient due to the drift field in the base layer (which is usually uniformly doped) because of the associated inhomogeneous doping. For a relatively small change in the concentration across the base layer (10^{17} in the p—n junction and 10^{18}-10^{19} cm^{-3} within the base) it is possible to produce an increase in efficiency and long-wave spectral sensitivity of silicon cells by introducing a drift field while maintaining at a reasonable level the diode parameters of the p—n junction and the lifetime of minority carriers in the base [84, 120,128-132].

The experimental realization of this model by slow diffusion of an impurity into the base wafer of a solar cell [128] was found to be laborious and time-consuming. The technology of exodiffusion of impurities in a vacuum from a base wafer doped in advance, which we have developed, was equally complex. The use of rapidly diffusing lithium [129] found practical application in the fabrication of space cells with radiation-resistant properties, not only because the drift field of sufficient intensity could be produced in the base layer by relatively simple technology, but also because lithium neutralized the recombination centers produced by radiation [22, 133,134]. Cheap terrestrial cells can be fabricated by growing an epitaxial silicon layer with graded impurity concentration on a single-crystal silicon wafer and then introducing the p—n junction (by thermal diffusion or by deposition of an epitaxial highly doped film with a different type of conductivity) from the side of the epitaxial layer (see, for example, [130-132]). The cost reduction is achieved because the epitaxial film is deposited on metallurgical silicon, which is a hundred times cheaper than semiconductor-type silicon [132]. The automatic doping of the epitaxial layer, while it is being grown by impurities from the substrate, produces the necessary impurity concentration gradient and, hence, the drift field [31]. Terrestrial measurements have shown that the efficiency of such cells is 12.2-13.5% [132] despite the use of defective epitaxial layers on metallurgical silicon, which is a serious fault in the manufacture of integrated circuits.

Solar cells with a strong drift field in the base were soon replaced by cells with a sharp isotype p—p$^+$ or n—n$^+$ junction on the metallic back contact, similar to the two-layer model considered in Section 2.1.

Near-intrinsic silicon could be used to produce high-efficiency solar cells by diffusing n- or p-type impurities from either side of the silicon wafer so as to produce the p—n junction at the necessary distance from the surface and, at the same time, achieve the optimum impurity gradient on the other side of the wafer [63]. Experience gained in the fabrication of n$^+$—p—p$^+$ or p$^+$—n—n$^+$ structures [77,135] shows that it is much simpler to produce a very thin isotype p—p$^+$ or n—n$^+$ junction on the metallic back contact than to establish a strong drift field, and that this is virtually as good from the point of view of minority-carrier collection from the base layer. The potential barrier on the isotype junction, produced by doping the base from the back, reflects the minority carriers from the back contact, increases their effective diffusion length, and reduces to practically zero the rate of surface recombination on the separation boundary between the base and the metallic back contact. There is also some reduction in the reverse saturation current of the cells. The doped back layer is produced by thermodiffusion, ion bombardment, or the implantation of aluminum (in the case of the p-type layer), followed by thermal treatment. The depth of the doped layer is usually between 0.2 and 0.5 μm, and the impurity distribution is practically the same as in the doped upper layer of solar cells.

The advantages of solar cells with isotype junction on the back surface become significant when the diffusion length of minority carriers in the base is greater than the thickness of the base layer or at least equal to it. This requirement means that the base layer must be a sufficiently pure semiconducting material of high enough resistivity, or the thickness of the base must be reduced to a value that is less than the diffusion length of minority carriers in the material. Figure 2.8 shows the diffusion length L and lifetime τ of carriers as functions of the resistivity ρ of the base silicon [136]. It can be used to choose the necessary base layer thickness with the necessary resistivity or, conversely, determine the resistivity for a given layer thickness (for a solar cell with efficient isotype junction on the back contact). However, such data can only be used for qualitative estimates. Accurate calculations must take into account the fact that the diffusion parameters and minority carriers depend not only on the resistivity but also on the type of conductivity of silicon, the method used to produce it, and the treatment to which it has been subjected. The necessary detailed data are given, for example, in [74]. The short-circuit current of a silicon solar cell is also given in [74] as a function of the ratio of the cell thickness l to the diffusion length L of minority carriers in the base and the efficiency of the cell under AM0 illumination as a function of L (Fig. 2.9) and τ for cells with and without the isotype junction on the back contact. The use of isotype junctions means that the base layer of high-efficiency solar cells can be made from silicon with very high values of lifetime and diffusion length of minority carriers, these being characteristic for silicon with near-intrinsic conductivity (the so-called i-conductivity).

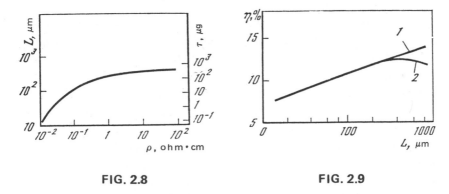

FIG. 2.8   **FIG. 2.9**

FIG. 2.8. Diffusion length and lifetime of minority carriers as functions of the base resistivity in silicon solar cells.

FIG. 2.9. Efficiency of silicon solar cells as a function of the diffusion length of minority carriers in the base under AM0 illumination: 1, 2) with and without the isotype barrier on the back surface, respectively.

Solar cells with the p–i–n or p$^+$–i–n$^+$ structure and their modifications [13,137] have exceptionally high sensitivity at long wavelengths. The current-voltage characteristic of such cells is nearly square because the high level of illumination under exposure to solar light ensures that the voltage drop across the base region is reduced to a minimum (the nonequilibrium carrier concentration produced under illumination of the high-resistivity base is much greater than the concentration of equilibrium carriers). The large initial diffusion length of minority carriers in the high-resistivity material extends the life of such cells in the Earth's radiation belts [22].

2.4 Solar cells transparent in the long-wave region beyond the fundamental absorption edge

The basic possibility of such solar cells is assured by the transparency of any pure high-resistivity semiconducting material beyond the fundamental absorption band (see Fig. 1.1). However, if the base of the solar cell is made from a relatively pure material with low-dopant concentration, the upper layer must be doped to a concentration that is practically equal to the solubility limit for the donor or acceptor impurity in the given semiconductor in order to reduce the spreading resistance of carriers separated by the p—n junction. Long-wave radiation will, of course, be strongly absorbed and reflected by this type of highly doped layer (see Figs. 1.2 and 1.3).

The low lifetime and diffusion length of minority carriers in the doped layer means that the thickness of the layer must be reduced to a value in the range 0.15-0.5 μm (see Chapter 1). Absorption of the infrared component of solar radiation (λ = 1.1-2.5 μm) by a cell with a doped layer of this thickness does not exceed 1-3% [23,111]. The tendency to reduce the depth of the p—n junction in modern solar cells has thus removed one of the obstacles to producing cells that are transparent in the long-wave region of the spectrum.

Two other obstacles, namely, absorption by the solid back contact and high reflection by the back surface of the element, were overcome by replacing the solid back contact with a grid and depositing an antireflective coating with an optical thickness of 0.3-0.4 μm [23,109]. Calculations show that, with a gridded back contact of a particular configuration, it is possible to preserve the series resistance and curve factor of the current-voltage characteristic of the transparent silicon solar cell practically at the level achieved for the conventional cell with a solid back contact [110,111]. Similar results have been obtained for solar cells made of gallium arsenide [23,108].

Transparent solar cells made of silicon and gallium arsenide were used in the first practical version of the cascade cell [108]. The equilibrium working temperature of transparent silicon solar cells used in space is much lower than that of conventional cells [23], so that the integrated absorption coefficient for solar radiation has been shown by direct measurement in space to amount to 0.72-0.73 rather than 0.92-0.93 (these are values characteristic for conventional cells with a solid back contact in the form of a fully reflecting metal) [110].

Silicon and gallium arsenide solar cells with a gridded back contact, which are transparent in the infrared beginning with λ = 1.1 μm, were produced in the USSR [23,108-111], whereas cells using Cu_2S-CdS thin-film structures were made in France [138]. Calculations have shown that the temperature of these cells in a geostationary orbit in space [138] should be lower by 10°C and the output power should be higher by 8-9%.

Nonphotoactive long-wave infrared radiation will not only be transmitted by the transparent solar cells, but will also be reflected by its back surface. This can be done by depositing a highly reflecting film of, say, aluminum, copper or silver on the back surface of the transparent solar cell free of the resistive current contact.

The reflector in the form of the usual titanium-palladium-silver three-layer structure can be deposited by evaporation in a high vacuum directly on the silicon surface, free from contact strips, or it can be produced simultaneously with the aluminum contact. However, the infrared reflection coefficient of this type of layer is reduced when the resistive aluminum contact is baked on at high temperature.

Figure 2.10 shows the spectral reflection coefficient of some of the high-efficiency solar cells with aluminum back contact used in [139]. Two of the cells (1 and 2) had polished back and front surfaces, whereas cell 3 had a nonreflecting black outer surface produced by selective etching which resulted in the appearance of closely spaced pyramidal etch pits. The surface of all three cells was covered with a film of tantalum pentoxide of different thickness. The baked-on reflecting aluminum contact could be used to increase reflection by the solar cells in the long-wave region beyond the fundamental absorption edge for $\lambda = 1.1$-2.5 μm to only 40%.

When the back contact is the three-layer titanium-palladium-silver deposit, the reflection coefficient in this region is no more than 20-30% [139], but can be increased to some extent by reducing the thickness of the titanium film [140]. Solar cells with black non-reflecting surface absorb practically completely not only radiation between 0.4 and 1.1 μm (the region of spectral sensitivity of the cell), but also the infrared radiation beyond the fundamental absorption edge. Transparent cells based on this principle cannot be produced [23].

It is much more convenient to increase reflection in nonphotoactive parts of the spectrum by using a highly reflecting metal deposited on the silicon surface in the openings of the grid contact on the back. The silicon surface can then be subjected to relatively slight heating (up to 150-200°C) to improve the adhesion of the layers while preserving at a high enough level the infrared reflection by the silicon-metal boundary [109]. The spectral reflection coefficient of the polished outer surface of silicon solar cells with a three-

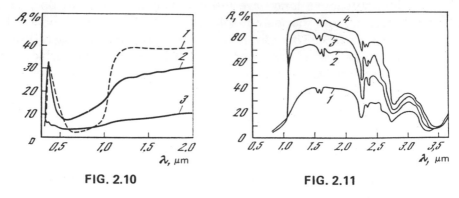

FIG. 2.10   **FIG. 2.11**

FIG. 2.10. Reflection coefficient of n—p-type silicon solar cells with an aluminum back contact and an antireflective coating of tantalum pentoxide deposited on the outer surface of the cell after the following treatment: 1, 2) polished; 3) black nonreflecting surface produced by selective etching.

FIG. 2.11. Reflection coefficient of a transparent silicon solar cell manufactured in the USSR with a three-layer coating on the polished outer surface and different reflecting layers on the back surface without contact strips (depth of p—n junction less than 0.5 μm): 1) nickel and titanium; 2) aluminum; 3) copper; 4) silver.

layer coating (antireflective film consisting of zinc sulfide + organosilicone adhesive + protective glass [23]) and different reflecting layers (copper, aluminum, silver, nickel, and titanium) on the back surface without contact strips is up to 75-95% despite the presence of selective absorption bands of the organosilicone compound between 1.1 and 1.5 μm (Fig. 2.11).

Comparably high reflection coefficients can also be attained by another simple method, namely, by using the organosilicone compound (on the back surface of transparent cells) to attach glass coated with aluminum or silver [141]. This procedure can be used to attach to the outer surface of the cell or group of cells a protective glass whose surface facing the element carries a grid of a reflecting metal at points lying above the current contacts to the individual solar cells or above the electrical junctions between them. By varying the width of the reflecting-grid lines, it is possible to adjust the temperature of such cells during an increase or a reduction in solar flux. Figure 2.12 illustrates the configuration of a module consisting

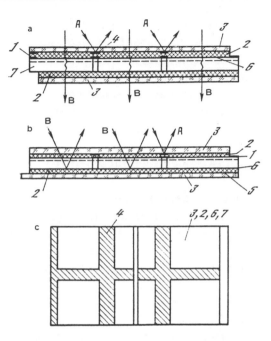

FIG. 2.12. Modules of parallel-connected transparent solar cells: a, b) with antireflective and reflective coating on the back surface, respectively; c) top view. 1) Current contacts and intercell connections; 2) transparent organosilicone compound; 3) protective glass cover; 4) grid of aluminum or silver reflecting strips above the upper current contacts and connections; 5) reflective coating on back glass; 6) antireflective coating; 7) solar cells; A) solar radiation; B) infrared solar radiation with $\lambda > 1.1$ μm and $\lambda > 0.9$ μm in the case of solar cells made from silicon and gallium arsenide, respectively.

of parallel-connected silicon cells, transparent in the infrared, with protective glass on both sides and a reflecting grid on the inner surface. Solar batteries consisting of such modules have a lower working temperature in space (by 25-35°C) and greater thermal stability. This was confirmed experimentally over long periods of time in space when the batteries were carried on board Venera-9 and Venera-10 automatic interplanetary stations [142].

It is important to note that the optical characteristics of transparent solar cells made from different semiconducting materials and carrying a reflecting back coating are very similar to the optical parameters of dichroic beam-dividing mirrors [143], which means that such cells may find important applications in high-efficiency photoelectric systems using spectral subdivision of solar radiation and subsequent conversion into electric power by cells with different spectral sensitivities. Transparent solar cells then perform two functions simultaneously, namely, beam-splitting and active conversion.

2.5 Solar cells with two-sided spectral sensitivity

Solar cells that can generate electric current when illuminated on both sides are useful both in space and in the laboratory because the use of such cells improves the limiting and working performance of semiconductor photoconverters of optical radiation into electric power. Solar batteries consisting of cells with two-sided sensitivity can convert (in low-lying orbits) not only direct solar radiation, but also radiation reflected by the Earth [143], since the albedo of the Earth can reach 0.8-0.9 over portions of the orbit (under continuous cloud cover). This means that appreciably greater power levels can be generated. In terrestrial applications, these solar batteries can be provided with additional reflectors that illuminate the back surface of the two-sided cells, or batteries can be mounted on tall supports, so that radiation reflected by snow or sand reaches the usually unilluminated back surface.

It would appear that the two-sided solar cell was first proposed in [144]. A doped n-type layer was produced on the front and back surfaces of silicon solar cells after the thermodiffusion of phosphorus into the boron-doped p-type silicon base. The current contacts on both surfaces were in the form of a grid, while the contact to the p-type base was a narrow strip running along one of the long sides of the cell. This was clearly insufficient to reduce the considerable spreading resistance to the flowing current which, in the case of the two-sided design (with a p—n junction on both sides of the cell) appeared not only in the doped region, but also in the base. This means, of course, that the output power per unit area of the two-sided cell could be increased (by a factor of 1.2-1.3 as compared with the conventional cell) only when the base region had a resistivity of 0.1-0.2 $\Omega\cdot$cm and a low spreading resistance [144]. Two-sided cells of this design, using silicon base wafers with resistivity

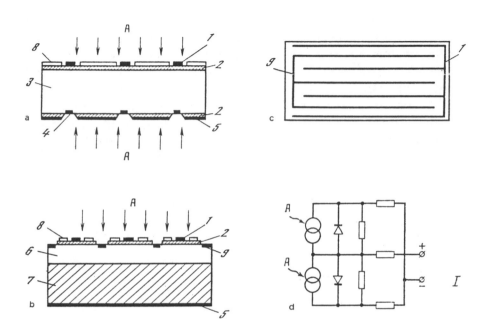

FIG. 2.13. Two configurations (a-c) and the equivalent electric circuit (d) of two-sided solar cells with two p—n junctions: a) on upper and back surfaces; b) on upper surface alone; c) top view of b; 1) upper current contact with the n$^+$ layer; 2) doped n$^+$ layer at 0.5 μm; 3) p-type base; 4, 9) contacts with the p-type base; 5) rear contact with the n-type layer; 6) p-type layer of depth 20 μm, produced by deep diffusion; 7) n-type base layer (thickness 200 μm); 8) antireflective coating; A) solar radiation.

of 0.5-1.0 $\Omega \cdot$cm, were found to be inferior under all illuminations as compared with conventional solar cells with a p—n junction on one side [144].

As noted in Section 2.4, the problem of reducing the spreading resistance in the base layer in transparent solar cells was solved by depositing a thin grid contact with optimized dimensions [23,111]. The same device was used in the theoretical paper [145] in the case

of two-sided silicon solar cells with p—n junctions on both sides.
Figure 2.13 shows the design of the two-sided solar cell, the equivalent electrical circuit, and the shape and disposition of the optimum contacts as suggested in [145]. The use of the comb-type
contact, not only on the highly doped upper and back surfaces but
also on the base layer, should undoubtedly improve the characteristics of the two-sided cells. This was confirmed by subsequent
calculations. The calculated collection coefficients of one-sided and
two-sided solar cells with different base-layer thickness l_b [145]
were used to calculate the short-circuit current I_{sc} (in % of the I_{sc}
of one-sided cells of conventional design with l_b = 400 μm) in the
case of two-sided cells under AM0 conditions as a function of the
base-layer thickness l_b for base resistivity of 10 $\Omega \cdot$cm. The results
were:

l_b, μm	10	15	20	35	50	100	200	400
Components of I_{sc} due to								
front junction	73.7	79.9	83.8	89.9	92.7	96.6	98.9	100.0
back junction	62.5	67.4	69.9	77.4	69.6	59.0	39.4	16.9
Resultant I_{sc}	136.2	147.3	153.7	167.3	162.3	155.6	138.3	116.9

Two-sided cells are better than one-sided cells for virtually all
values of the base-layer thickness. For an optimum thickness of
20-100 μm (base with ρ = 10 $\Omega \cdot$cm), the increase in I_{sc} (and, consequently, in the output power for a high curve factor) exceeds 50%
as compared with I_{sc} for one-sided cells with base-layer thickness
of 400 μm.

Two-sided solar cells were subsequently constructed in different
countries and measurements of their optical and electrical characteristics generally confirmed the calculations reported in [145].

The fabrication of the contact grid for two-sided solar cells with
two p—n junctions requires the use of double photolithography with
the imposition of intermediate templates, which introduces an appreciable complication into the solar cell production technology. Moreover, the greater area occupied by the p—n junction gives rise to a
higher reverse saturation current and a lower shunt resistance as
compared with conventional single-junction cells.

It has been suggested that two-sided solar cells could be made by combining elements that are transparent in the infrared with elements incorporating an isotype junction on the back surface with the n^+–p–p^+ or p^+–p–n^+ structure (Fig. 2.14) [5,146,147]. When the base-layer thickness is reduced (or the diffusion length of minority carriers is increased by using higher resistivity silicon), such two-sided cells are just as efficient in the conversion of light incident from the rear as cells with a second p–n junction on the back surface. Calculated and experimental characteristics of the new version of the two-sided solar cell are shown in Figs. 2.15 and 2.16. They were obtained for different rates of surface recombination (S_b and S_d) and different hole and electron diffusion coefficients (D_b and D_d) in the base and doped layers, respectively [146,147].

The introduction of the isotype junction into the transparent solar cell produces a sharp reduction in the rate of surface recombination S on the back surface and an increase in the minority-carrier collection coefficient (for $L/l > 1$) of two-sided solar cells illuminated from the back, up to values typical for the situation where the solar cell is illuminated from the front surface.

In contrast to the two-sided cells with two p–n junctions, the reverse saturation current in two-sided cells with an isotype junction does not increase when only the upper, front surface of the cell is illuminated. At the same time, the deposition of current contacts on both surfaces can be performed simultaneously by a single photolithographic process (simultaneous exposure of both sides).

Two-sided solar cells are no more difficult to fabricate than the solar cells and batteries that have been in practical use in space for many years [142]. The isotype barrier above the gridded back contact can be introduced by bombardment with boron ions, followed by thermal annealing [146] or the deposition of a transparent conducting SnO_2 film produced by chemical pulverization (the isotype junction is probably established as a result of the influence of the built-in electric charge [147]).

To increase the efficiency of two-sided solar cells with the isotype back junction, it is desirable to use a base layer with higher resistivity as compared with the usual material [146,147], for example, silicon with $\rho = 7.5$-10 $\Omega \cdot$cm instead of the single-crystal silicon with $\rho = 0.5$-1.5 $\Omega \cdot$cm (or reduce the base layer thickness). This is clearly indicated, for example, by the results reported in

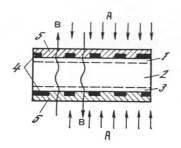

FIG. 2.14

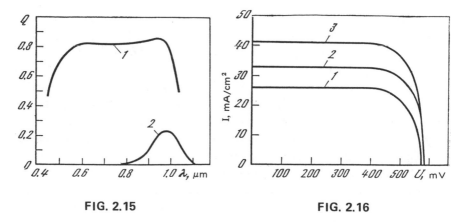

FIG. 2.15   **FIG. 2.16**

FIG. 2.14. Transparent two-sided solar cell with n^+-p-p^+ or p^+-p-n^+ configuration: 1-3) n^+-, p-, and p^+-type layers in the case of the n^+-p-p^+ structure and p^+-, p-, and n^+-type layers in the case of the p^+-p-n^+ structure, respectively; 4) current contacts; 5) antireflective and protective coatings; A) solar radiation; B) infrared component of solar radiation passing through the cell.

FIG. 2.15. Measured collection coefficient of a transparent silicon solar cell with the n^+-p-p^+ (1) and n^+-p (2) structure illuminated from the back (l_d = 0.5 μm, both sides; $L_{p^+} = L_{n^+}$ = 0.5 μm; $S_{n,p}/D_{p,n}$ = 10^5 cm^{-1}) for different dimensions of the base layer: 1) S_n/D_n = 0, L_n = 450 μm, l_b = 300 μm; 2) S_n/D_n = 10^5 cm^{-1}, L_n = = 100 μm, l = 250 μm.

FIG. 2.16. Current-voltage characteristics of a transparent two-sided solar cell with the n^+-p-p^+ structure (structure parameters are indicated in Fig. 2.15), measured under the AM0 simulator producing an irradiance E at a temperature of 25°C: 1) only rear side illuminated (E = 1360 W/m^2); 2) only front side illuminated (E = 1360 W/m^2); 3) simultaneous illumination of front and back surfaces (E = 1360 and 420 W/m^2, respectively).

[5,146]. Determinations of the minority-carrier diffusion length in the base of such cells are discussed in [92].

Experiments have shown that, when cells of the usual transparent design (see Fig. 2.12) are illuminated from the back, the increase in the current and output power is no more than 10 or 20% of the initial values (this was confirmed, among other things, during the first few hours of the Venera-9 and Venera-10 missions), and the main advantage of transparent cells of conventional design is that the working temperature of the solar batteries is lower [142]. When the thermal conditions in two-sided solar cells are calculated for space and terrestrial operation, it is convenient to make use of the nomograms and tabulations given in [148].

The use of two-sided cells with isotype junctions on the back surface in low-orbit satellites results in the availability of reserves of power [143]. In this experiment, the solar irradiance intercepted by the back face of the two-sided cell was 0.3 of the irradiance on the upper, front surface because the average albedo of the Earth is close to this figure. Consequently, these measurements can be used to estimate the possible gain in power generated by solar batteries consisting of two-sided cells with isotype junction on the back surface when they are mounted on low-orbit satellites (orbit altitudes of 200-400 km). These results were subsequently confirmed qualitatively by a direct experiment in space [149]. The average albedo of the Earth during this flight was 0.25, and the current drawn from the two-sided solar batteries was, on average, 17-18% greater (15 ± ± 2% during the first ten orbits) than that delivered by one-sided solar batteries of conventional design.

We note in conclusion that, in contrast to solar cells of the usual design (see Chapter 1), the current, efficiency, and output power of two-sided solar cells of either design (with the second p—n junction or the isotype barrier on the back surface) are greater when the base-layer thickness is reduced to 35-50 μm, which means that these cells may be useful whenever it is necessary to improve the specific parameters of solar cells, for example, the power-to-weight ratio.

2.6   Composite solar cells based on homojunctions
and heterojunctions in gallium arsenide

In the mid 1950's gallium arsenide attracted the attention of a number of researchers because solar cells made of this semiconductor, with the p—n junction in the homogeneous material, were found to have efficiencies of conversion of solar radiation into electric power comparable to that of mid 1950 solar cells (n = 4-6%) [150, 151]. The p—n junction was produced by diffusing p-type cadmium (later, zinc) into the initial n-type wafers.

Despite some disadvantages (brittleness and high density), gallium arsenide has undoubted advantages as compared with silicon. Because of its wide band gap, gallium arsenide has a limited ability to convert long-wave solar radiation (it absorbs radiation with wavelenghts below 0.9 μm). However, the same feature leads to much lower reverse saturation current, namely, $I_0 = 10^{-9}$-10^{-10} A/cm^2 (for silicon solar cells, $I_0 = 10^{-6}$-10^{-7} A/cm^2) [69,70] which, in turn, results in greater open-circuit voltage (0.7-0.8 V for a p—n junction in a homogeneous material) and reasonable efficiency (10-12%) has been recorded under a sun simulator. The same properties of this semiconducting material ensure that its efficiency falls more slowly with increasing temperature, namely, at a rate of 0.25%/°C (which may be compared with 0.45-0.46%/°C for silicon cells).

Gallium arsenide cells were used in the solar batteries serving as the source of power on the Soviet interplanetary stations Lunokhod-1 and Lunokhod-2 [142]. The batteries were running for over ten months, and the results obtained confirmed the above advantages of gallium arsenide. The batteries were operated at 130-140°C on the lunar surface and generated nearly twice as much electric power as the figure calculated for a silicon solar cell under comparable conditions [142]. In this case, it was particularly important to have highly efficient batteries because the space available for them was strictly limited (the batteries were mounted on the hinged door of the spacecraft). The radiator of the spacecraft was covered by heat-reflecting radiation-resistant glass coated with aluminum or silver on the inner surface [23]. The ratio of the integrated solar absorption coefficient α_S to the integrated intrinsic thermal emission coefficient ϵ of this cover was less than 0.2 [23].

At the beginning and end of the lunar day, when the Sun was low above the lunar horizon, the solar radiation reflected by the radiator was able to reach the open door of the spacecraft. Telemetry data then indicated that there was a clear rise in the current delivered by the solar batteries, and their temperature rose from 120 to 140°C [142]. The radiator was thus also used as a kind of concentrating reflector.

These solar cells, incorporating the p—n junction in homogeneous gallium arsenide, are conveniently mounted on spacecraft launched toward the Sun or to the more distant planets of the solar system [19,106]. As the photocurrent rises under increasing incident flux density (for example, as the spacecraft approaches Venus or Mercury) and the initial reverse saturation current I_0 decreases, the power temperature gradient of these cells may amount to 0.15%/°C, which is lower by a factor of three than that for silicon solar cells of conventional design. Gallium arsenide solar cells are effective under low illumination (not only in space but also on the ground, for example, as highly sensitive light meters in photography and small power sources for clocks and calculators). This is also due to the low values of I_0, the steep current-irradiance characteristic, and the rapid rise of U_{oc} and the load voltage when the irradiance changes by a small amount at low irradiance levels (between a few units and a few tens of lux).

We note that the most important ways of improving the performance of solar cells incorporating the p—n junction in homogeneous silicon are practically the same as those in the case of gallium arsenide. Their performance can be improved by further reducing the depth of the p—n junction down to 0.1-0.2 μm, by increasing the minority-carrier diffusion length in the base, by using built-in electrostatic drift fields and additional isotype barriers and p—n junctions, and by optimizing the contact system, especially when concentrated radiation produced by parabolic reflectors or flat Fresnel lenses is employed. The figure η = 21.1% has been attained for a solar cell consisting of homogeneous gallium arsenide and a shallow p—n junction, illuminated by AM1 solar radiation with a concentration ratio of 24, while the corresponding result for the concentration ratio of 325 was η = 16.9% [152]. When exposed to the AM1 spectrum at a temperature of 80°C, these cells were found to have η = 15.4% and U_{oc} = 0.97 V.

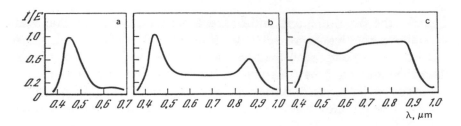

FIG. 2.17. Wavelength dependence of the sensitivity of solar cells with the GaP—GaAs heterojunction and a graded structure between GaP and GaAs when the p—n junction lies in GaP (a) and at different depths in the graded-gap structure (b, c).

The elements forming the semiconducting gallium arsenide are also found in a number of binary, ternary, and quaternary semiconducting compounds [153] with a lattice constant similar to that of gallium arsenide but different band gap, depending on the chemical composition. This offers us the opportunity of forming on the surface of the homogeneous gallium arsenide a layer of another semiconducting compound. This is done by producing a heterojunction in which, since the lattice constants of the conducting materials are very similar, there are no mechanical stresses or recombination centers. At the same time, by gradually varying the chemical composition and therefore the band gap E_g within the heterojunction, it is possible to produce the so-called graded structure, for example, a structure in which E_g is high on the surface and decreases with depth. This is the optimum case for cells used to transform solar radiation [63,84] because photons corresponding to the ultraviolet or short-wave visible radiation are absorbed in the uppermost layers of the solar cells.

These heterostructures are used not only to modify (usually to expand) the spectral sensitivity range, but also to introduce high electrostatic drift fields into the gallium arsenide cell. These fields are due to the dopant gradient within the cell (which is the only possible way of producing the drift field in silicon solar cells) and the band gap gradient.

One of the simplest and most original technological methods of producing this graded structure on the surface of the gallium arsenide solar cell is discussed in [154]. The necessary structures were produced by employing the well-established technique of thermal diffusion (instead of liquid or gas epitaxy). An n-type gallium arsenide wafer [with $N_n = 10^{17} \text{-} 5 \cdot 10^{17} \text{cm}^{-3}$ and $\mu_n = 3000$ $\text{cm}^2/(\text{V} \cdot \text{sec})$] was used to produce the solar cell. Thermodiffusion of phosphorus in an evacuated quartz ampoule (residual pressure 10^{-6} mm Hg) was used at a temperature above $900°C$ to produce a surface layer on gallium phosphide and a thin transition region whose composition varied gradually from GaP to GaAs. This corresponded to a variation of the band gap from 2.25 to 1.35 eV (at room temperature). The combined thickness of the GaP layer and the graded-gap region was 5-7 μm. Subsequent thermodiffusion of the acceptor zinc impurity into these structures was used in a similar way to produce the p—n junctions, whose depths could be varied by varying the thermodiffusion parameters.

The spectral sensitivity of these cells is illustrated in Fig. 2.17. The differences between the curves are due to differences in the depth of the p—n junction. Thus, the curve in Fig. 2.17a has a well-defined maximum at 0.45 μm, due to the p—n junction lying in the GaP surface layer. In Fig. 2.17b and c, the two maxima (at 0.45 and 0.85 μm) correspond to the p—n junction lying in the graded-gap region between the phosphide and the gallium arsenide. Consequently, the spectral sensitivity of these solar cells can be varied as desired in the wavelength range between 0.45 and 0.85 μm. When the thermodiffusion length of zinc is large, and the p—n junction lies in pure gallium arsenide, the spectral sensitivity curve consists practically of the single maximum at 0.85 μm (this case is not shown in Fig. 2.17). The open-circuit voltage U_{oc} of solar cells based on the gallium phospide—gallium arsenide heterostructure was found to be up to 0.8 V, but the efficiency did not exceed 4-5%.

It was subsequently discovered that, since the lattice constants were almost completely matched, the heterojunction in the system consisting of the solid solution of aluminum in gallium arsenide and gallium arsenide had a very low density of states and a low density of recombination centers on the separation boundary [155], so that these structures exhibited two-sided carrier collection with a high quantum yield. The heterostructure was used to produce a solar cell with $\eta = 11\%$ under exposure to the AM0 simulator [116].

The most widely used solar cells are those based on the hetero-systems p-Ga$_{1-x}$Al$_x$As—p-GaAs—n-GaAs, produced by liquid or gas epitaxy with simultaneous thermodiffusion of the acceptor zinc impurity. The main p—n junction separating the carriers is then located in the gallium arsenide base, and the layer consisting of the solid solution of aluminum in gallium arsenide acts as a wide-gap filtering window so that, owing to the presence of the isotype p—p junction on the gallium arsenide surface, losses by surface recombination are almost totally avoided [116,117,156-164].

The thickness and chemical composition of the wide-gap filter may vary, and this may have a considerable effect on the properties of the resulting solar cells. For example, an increase in the thickness of this layer and the p-GaAs layer, and an increase in the dopant concentration in the two layers, are accompanied by a sharp reduction in the series resistance of the cells (the cells are then conveniently used with high degrees of concentration of the solar flux) [160]. When the thickness of the upper layers of the cell is reduced, optical losses by absorption in these layers are practically totally eliminated [158]. By varying the chemical composition of the window (in particular, its aluminum concentration), it is possible to produce a graded-gap surface structure that helps in the collection of carriers generated by short-wave light in the upper layers of the solar cells. The optical and electrical properties of solar cells incorporating these heterostructures can be calculated and optimized if we know the optical constants of the semiconducting layers, the band gap, and the nature of optical transitions in the fundamental absorption band as functions of the composition of the material.

The optical constants of single-crystal and thin-film gallium arsenide are listed in [28,29,158], while those of the solid solution of aluminum and gallium arsenide are given in [158-160]. The dependence of the band gap of this compound on its composition is reported in [165], where it is noted that the semiconductor $Al_xGa_{1-x}As$ for $x \leqslant 4$ is characterized by direct optical transitions whereas, for $0.4 \leqslant x \leqslant 0.8$ (chemical stability limit), the wavelength dependence of the absorption coefficient has a form typical for indirect transitions. This means that, even before the cells are developed, it is possible to reduce absorption in the solid solution by using a thin-layer window with a high value of x or a graded structure of small thickness (preferably low x on the solid solution—air boundary and high x on the solid solution—gallium arsenide separation boundary). It is important to note at this point that, in

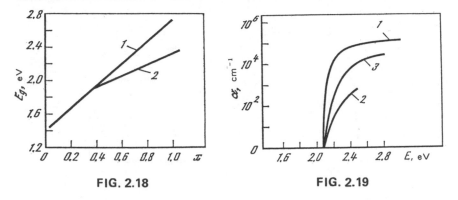

FIG. 2.18   FIG. 2.19

FIG. 2.18. Gap width of the semiconducting compound $Al_xGa_{1-x}As$ as a function of its composition x for different types of optical transition: 1) direct; 2) mixed.

FIG. 2.19. Wavelength dependence of the absorption coefficient of $Al_xGa_{1-x}As$ for direct (1) and indirect (2) optical transitions and for the compound with x = 0.86 (3).

the case of sufficiently thick layers of solid solutions, which are technologically relatively easily produced and have advantages from the point of view of deposition of a reliable electrical contact, it is useful to employ solid solutions with the reverse dependence of composition on depth, or simply uniform layers with relatively high values of x [160].

Figure 2.18 shows the band gap of $Al_xGa_{1-x}As$ as a function of the composition of the semiconducting compound for direct optical transitions (curve 1) and for the mixed model consisting of direct (x ≤ 0.4) and indirect (x ≥ 0.4, curve 2) transitions [160, 165]. Figure 2.19 shows the wavelength dependence of the absorption coefficient of $Al_xGa_{1-x}As$ for direct and indirect transitions (curves 1 and 2, respectively). These data were used in [160] to calculate the collection coefficient. The figure also shows (curve 3) the analogous dependence for x = 0.86, proposed in [158] for the mixed model.

The influence of the thickness and composition of the upper layers on the optical characteristics and efficiency of solar cells with a wide-gap $Al_xGa_{1-x}As$ window and a p—n junction under it in single-crystal gallium arsenide can be investigated by comparing the results reported by different researchers. Figure 2.20 shows the

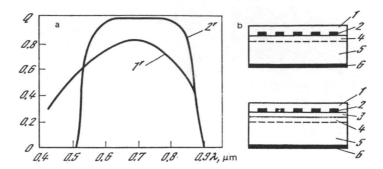

FIG. 2.20. Wavelength dependence of the collection coefficient (a) of gallium arsenide cells with a homogeneous p—n junction (1′) and the heterojunction p-Ga$_{0.3}$Al$_{0.7}$As—p-GaAs—n-GaAs (2′), and the arrangement scheme of the cell layers of both types (b): 1) antireflection and protective coatings; 2) top current contact; 3) widegap filter-window of p-Al$_x$Ga$_{1-x}$As; 4) p-GaAs; 5) n-GaAs base, 250-300 μm; 6) back contact.

measured wavelength dependence of the carrier collection coefficient of a gallium arsenide solar cell of conventional design with a homogeneous p—n junction (curve 1′) and a heterostructure on the surface (curve 2′) [117]. The cell with the heterostructure had the following composition and layer thickness: p-Ga$_{0.3}$Al$_{0.7}$As (8 μm), p-GaAs (0.7 μm), and n-GaAs (300 μm). These solar cells were found to have an efficiency in excess of 20% under terrestrial conditions. At the same time, because of the relatively large thickness of the graded-gap filter-window and the low series resistance, these cells are useful for measuring enhanced solar flux densities. However, the short-wave edge of the spectral sensitivity of these solar cells lies at 0.51-0.52 μm for the same reasons (Fig. 2.20).

Numerical optimizations of solar-cell parameters for gallium arsenide cells with heterojunctions [158,160,166] have shown that, by reducing the thickness of the top layer of the solid solution and by varying its composition (increasing the aluminum concentration), it is possible to extend quite substantially the spectral sensitivity range of these cells toward the short-wave part of the spectrum. The calculated wavelength dependence of the collection coefficient of these solar cells is plotted in Fig. 2.21 [166] for different thicknesses of the Al$_{0.86}$Ga$_{0.14}$As layer and the following cell parameters: p-GaAs and n-GaAs layer thickness 1.5 μm and 250 μm,

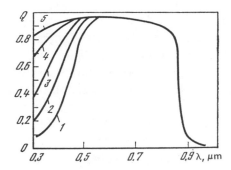

FIG. 2.21. Wavelength dependence of the collection coefficient of solar cells with the structure p-$Al_{0.86}Ga_{0.14}As$–p-GaAs ($l =$ = 1.5 μm) –n-GaAs (l = 250 μm) for the following thicknesses of the top layer of the solid solution: 1) 1.0 μm; 2) 0.5 μm; 3) 0.25 μm; 4) 0.1 μm; 5) 0.05 μm.

respectively, carrier concentration in all layers 10^{18}-$3 \cdot 10^{18}$ cm^{-3}, diffusion length in the top layer of the solid solution L_n = 0.5 μm, carrier mobility μ_n = 250 cm^2/(V·sec); the corresponding figures for p-GaAs were: L_n = 5 μm, μ_n = 2500 cm^2/(V·sec), while in n-GaAs, the parameters were L_p = 0.5 μm, μ_p = 150 cm^2/(V·sec).

It is important to note that the high values of the collection coefficient in the long-wave region (λ = 0.6-0.9 μm) are due to the relatively large carrier diffusion length in p-GaAs (greater by a factor of three than the layer thickness).

Experiment confirms these calculations. The current-voltage load characteristic of prototype solar cells [167] shows that their efficiency under AM1 is appreciably greater than 20% and may even be as high as 25% if, for example, one uses very thin top layers of the solid solution with a graded chemical composition and graded gap width [166,168].

2.7 Thin-film solar cells

In cells made from semiconducting materials with direct optical transitions, the conversion of solar energy into electric power occurs in the surface layer of depth no more than 10 μm, and most of the incident radiation is absorbed in a layer of only 2-3 μm (see, for example, Fig. 1.1, curve 2 and Fig. 1.24b). The initial cell material is usually in the form of 250-300 μm base wafers (see Fig. 2.20b), which cannot be justified from the standpoint of physics or economics. This situation can only be explained by the great technological difficulties and high rates of rejection in manufacture whenever one tries to reduce the thickness of the brittle base wafers of single-crystal gallium arsenide and direct-gap semiconductors. Many researchers have therefore tried to produce thin films of semiconducting materials with direct optical transitions as a way of developing more efficient and cheaper solar cells.

Thin-film solar cells based on the heterostructure
copper sulfide–cadmium sulfide

Heterostructures involving thin films of $A^{II}B^{VI}$ compounds, especially cadmium sulfide, occupy the central position among thin-film solar cells [169,170]. In the very first solar cells made from this semiconducting material [171], the separating barrier was produced by depositing semitransparent layers of silver, copper, gold, or platinum on the cadmium sulfide surface. Practically all the subsequent solar cells made use of the copper sulfide–cadmium sulfide heterojunction, where the copper sulfide was produced by substituting copper for cadmium atoms in the course of the chemical reaction (at 90-95°C) between cadmium sulfide and cuprous chloride in liquid [172,173] and solid [174] phases. In the latter case, cuprous chloride was first deposited on the surface of cadmium sulfide films evaporated in a vacuum.

The first is the so-called wet method. When it is used, the surface of the solar cell and of the heterojunction itself has numerous pits and projecting grains, and the surface structure is enhanced by chemical etching. This reduces the reflection coefficient of the cell surface, but increases the reverse saturation current.

The second method — the so-called dry method — results in an almost planar heterojunction, plane-parallel relative to the substrate, but the photosensitivity of the copper sulfide films produced in the course of the reaction in the solid phase is somewhat lower than that of films produced by the wet method.

Thin-film solar cells based on the copper sulfide–cadmium sulfide heterosystem are usually divided into two types, namely, back barrier and front barrier [13,19].

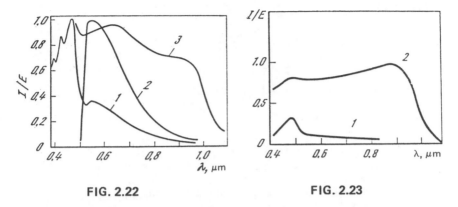

FIG. 2.22   FIG. 2.23

FIG. 2.22. Wavelength dependence of the sensitivity of front-(1, 3) and back-barrier (2) solar cells based on the copper sulfide—cadmium sulfide heterosystem for different values of the copper-sulfide thickness: 1, 2) hundreds of Ångstroms; 3) thousands of Ångstroms.

FIG. 2.23. Wavelength dependence of the sensitivity of front-barrier solar cell based on the copper sulfide—cadmium sulfide heterosystem with deposited copper current contacts: 1, 2) before and after thermal treatment, respectively.

In the front-barrier design, the cadmium sulfide film is deposited in a vacuum on a substrate in the form of a molybdenum foil, a polyimide film, or a zinc-coated copper foil heated to 200-300°C. The copper sulfide film is produced by either the dry or wet method. The contact with the layer is in the form of a grid of copper strips, evaporated in a vacuum through a stencil mask, or a gold-plated copper grid attached with a conducting paste (it can also be held in position by the sticky film of protective polymer film).

When the back-barrier solar cells are fabricated, the cadmium sulfide is deposited in the same way on a heated glass cover slip or a glass plate coated with transparent conducting SnO_2 and In_2O_3 (ITO) or Cd_2SnO_4 [175], and the copper sulfide—cadmium sulfide heterojunction is produced in the same way as before. The contact with the copper sulfide layer can be in the form of a solid copper layer, produced by evaporation, since the back-barrier thin-film cell is illuminated on the glass side.

The thickness of the cadmium sulfide layer is usually between 2 and 40 μm and that of the copper sulfide layer between 0.05 and

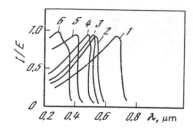

FIG. 2.24. Spectral sensitivity of front-barrier solar cells produced by vacuum evaporation of copper selenide (1) and copper sulfide (2-6) layers onto the base for different semiconducting compounds: 1, 2) CdS; 3) $Zn_{0.1}Cd_{0.9}S$; 4) $Zn_{0.15}Cd_{0.85}S$; 5) $Zn_{0.4}Cd_{0.6}S$; 6) ZnS.

0.15 μm. The band gap of copper sulfide is 1.2 eV and that of cadmium sulfide 2.4 eV [176]. The back-barrier cell is insensitive (Fig. 2.22, curve 2) at short wavelengths; the top layer of cadmium sulfide that faces the incident light then acts as a filter-window which absorbs practically all radiation below 0.5 μm. The spectral sensitivities of the front-barrier solar cell, based on the copper sulfide—cadmium sulfide heterosystem (curve 1), and of the back-barrier cell (curve 2), were obtained for cells prepared by the wet method with a copper sulfide layer of a few hundred Ångstroms [177]. When the copper sulfide layer thickness is increased to 0.15 μm, the spectral sensitivity of the front-barrier cell is found to rise rapidly in the long-wave part of the spectrum [138,172,178] (Fig. 2.22, curve 3). Figure 2.23 shows the effect of thermal treatment (applied after the fabrication and deposition of the copper contact strips on the working surface) on the spectral sensitivity of front-barrier solar cells based on the copper sulfide—cadmium sulfide heterosystem [179].

It is probable that the diffusion of copper atoms from the contacts to the surface layer, which occurs during the thermal treatment, improves both the stoichiometric composition of the copper sulfide layer and its photosensitivity. The position of the long-wave sensitivity edge prior to the thermal treatment (Fig. 2.23, curve 1) corresponds to the absorption edge of cadmium sulfide (E_g = 2.4 eV). This suggests that the short-wave part of the sensitivity of front-barrier cells based on the copper sulfide—cadmium sulfide heterosystem is due to the cadmium sulfide, whereas the sensitivity at other wavelengths is due to the copper sulfide. A similar enhancement of the effect of thermal treatment after the preliminary deposition of a semitransparent copper film on the cell surface was also reported in [138].

This type of solar cell and its band diagram, based on the virtually complete absorption of light in the copper sulfide, is described in [176,180]. The minority carriers (electrons) produced by light in the degenerate p-Cu_2S layer (copper vacancies play the role of acceptors, the hole concentration being 10^{19} cm^{-3}) diffuse through the heterojunction or are transported by the space-charge field into the cadmium sulfide. The dominant effect of the copper sulfide on the photocurrent generated by these solar cells was confirmed by experiments in which this layer was gradually removed by etching, and this was found to result in a sharp reduction in sensitivity at long wavelengths.

Copper sulfide films deposited in vacuum onto cadmium sulfide layers have practically no photosensitivity prior to the thermal treatment and act only as transparent current contacts to the heterojunction [181]. The cell sensitivity is then practically entirely determined by the composition and properties of the base layer, which can be varied, for example, by adding to the cadmium sulfide a certain amount of wide-gap semiconducting material such as zinc sulfide (E_g = 3.66 eV) [170]. This property of vacuum-deposited copper sulfide layers has been exploited in producing variable-sensitivity detectors of ultraviolet and visible radiation (Fig. 2.24) [181]. In many ways, the composition of the copper sulfide layer determines the short-circuit current of thin-film solar cells based on the copper sulfide—cadmium sulfide heterosystem [138]. The measured dependence of the short-circuit current of these solar cells on copper concentration in the copper sulfide layer (the quantity x in the formula Cu_xS) and on the structural composition of the compound (Fig. 2.25) [138] shows that high-quality solar cells of this kind can be obtained by ensuring that the composition of the copper sulfide layer deposited on the cadmium sulfide surface is as close as possible to the composition of chalcocite.

A similar dependence on x in the formula Cu_xS was put forward in [182] for the open-circuit voltage of solar cells based on the copper sulfide—cadmium sulfide system.

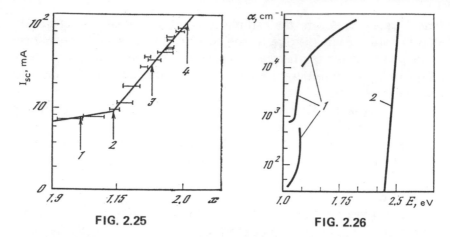

FIG. 2.25   FIG. 2.26

FIG. 2.25. Short-circuit current drawn from a front-barrier thin-film cell based on the copper sulfide—cadmium sulfide hetero-system as a function of the concentration of copper in copper sulfide for different structural compositions of this layer: 1) djurleite + digenite; 2) djurleite; 3) chalcocite + djurleite; 4) chalcocite.

FIG. 2.26. Wavelength dependence of the absorption coefficient of copper sulfide (1) and cadmium sulfide (2).

Because of this well-defined effect of the copper sulfide layer on the properties of thin-film solar cells of this design, much effort has been devoted to studying the optical properties of copper sulfide. This is essential for the calculation of the collection coefficient and the efficiency of solar cells, and for choosing the optimum antireflective and protective coatings. Figure 2.26 shows a summary [173,183] of the most reliable data on the wavelength dependence of the absorption coefficient of copper sulfide and cadmium sulfide.

The wavelength dependence of the other optical constants, namely, the refractive index and the absorption coefficient (n and k, respectively) of copper sulfide, was calculated from the reflection and transmission data for thin layers (l = 625-650 Å) of this compound for the wavelength sensitivity range of solar cells based on the copper sulfide—cadmium sulfide heterosystem [23]:

λ, μm	0.4	0.5	0.6	0.7	0.8	0.9	1.0
n	1.0	1.3	2.2	2.3	2.2	2.1	2.0
k	1.6	0.8	0.2	0.4	0.6	0.9	1.4

These results are very similar to those reported in [184] for digenite crystals. The optical constants of cadmium sulfide and other traditional semiconducting materials are listed in [25] and in a number of other sources, for example, [185].

The thin-film solar cells based on the copper sulfide—cadmium sulfide heterosystem and fabricated in different countries have AM1 efficiencies of 4-7% [177-183], but efficiencies of up to 10% have already been attained under the AM1 simulator [186]. The higher efficiency was achieved by introducing the improvements suggested above and, in particular, by using a combination of contacts deposited on copper sulfide and a contact grid attached with a conducting paste. This sharply reduced the series resistance of these cells [176, 183]. Moreover, the copper sulfide layer was produced by the wet method (and not the dry method, which increases U_{oc} to 0.58 V and gives rise to higher reflection losses), but the concentration of hydrochloric acid in the etching solution applied to the cadmium sulfide surface before the application of copper monochloride was reduced from 55 to 25%, and the etching time was increased from 2 to 20-40 sec at 60°C. As a result, the pyramidal etch pits on the surface of the final cells amounted to no more than 1 μm, so that reflection losses were sharply reduced. The short-circuit current was 22.2-24.7 mA/cm^2, while U_{oc} was maintained at 0.54-0.58 V for a high fill factor of the current-voltage characteristic.

The efficiency of thin-film cells can be increased further by using the structures that were found to be effective in improving the characteristics of silicon and gallium arsenide solar cells. In particular, additional doping of the copper sulfide surface with copper atoms can be used to produce a p^+—p structure in the upper layer, while the doping of cadmium sulfide with zinc, cadmium, or aluminum produces a two-layer n—n$^+$ structure in the base layer. By adding zinc sulfide to the cadmium sulfide, it is possible to produce a graded structure in the base, and reduce the difference between the lattice constants of the semiconducting materials forming the heterojunction [176,178,181,183,187]. When the eventual cell is subjected to thermal treatment in air at 200°C, a layer of the wide-band semiconductor $Cu_x S_y O_{1-y}$ may form on the surface and may play the same role as the $Al_x Ga_{1-x} As$ layer in gallium arsenide cells with heterojunctions [188].

However, there are ways of improving the parameters that are characteristic for and specific to these particular types of solar cell. For example, replacement of the Cu_2S layer with InP or $CuInSe_2$ [189,190] leads to a substantial rise in the collection coefficient, a reduction in the density of states near the heterojunction (the lattice constants of cadmium sulfide and these materials are very similar), and a substantial improvement in the stability of the characteristics of thin-film cells in time, when chromium telluride is used instead of copper sulfide [191] although, of course, the life of these cells under prolonged operation is significantly extended by using multilayer antireflective and protective coatings [23]. To improve the rate of deposition of the cadmium sulfide layer, and to reduce costs, attempts have been made to replace evaporation in a quasiclosed volume by chemical pulverization in air [192], or by using gas transport reactions [193].

The electrical and optical properties of many heterosystems based on $A^{II}B^{VI}$ compounds suggested for thin-film solar cells, for example, p-ZnTe—n-CdSe, p-ZnTe—n-CdTe, p-CdTe—n-CdS, p-CdTe—n-ZnSe, p-CdTe—n-CdZnS, and so on, have been described in some detail, for example, in [169,170]. The efficiency of solar cells based on these systems is, as yet, lower than that achieved for the copper sulfide—cadmium sulfide heterosystem, but some of them, for example, p-CdTe—n-CdS solar cells, have attracted attention because of their low temperature coefficient of power and their stable characteristics [194].

There are plans for the large-scale utilization of thin-film solar cells based on the copper sulfide—cadmium sulfide heterosystem (and their modifications) in terrestrial solar power engineering but, at present, these cells are mostly used as small and very sensitive detectors of the solar ultraviolet and of radiation produced by artificial sources [178,181]. Tests performed on these cells in space [23] suggest that they can be used as a basis for developing flexible solar batteries of moderate efficiency (as compared with the efficiency of batteries consisting of single-crystal silicon or gallium arsenide solar cells), but with a relatively high power-to-weight ratio.

Thin-film solar cells based on amorphous silicon

A large number of recent studies have been devoted to thin-film solar cells based on amorphous silicon (the so-called α-Si) — an interesting semiconducting material produced mainly through the decomposition of silicon compounds in a high-frequency discharge in vacuum. A reasonably complete review of the development of solar cells based on α-silicon, which began to develop intensively in 1969-1970, is given in [195]. Early studies showed that the number of states and recombination centers in the band gap of amorphous silicon produced in a discharge is lower by several orders of magnitude than in silicon deposited on different substrates by evaporation in vacuum. The properties of amorphous silicon are improved further by including 5-50 at.% of nitrogen in the material. This produces what is fact an alloy of silicon and hydrogen which, in turn, facilitates the doping of the medium with phosphorus or boron, and hence the appearance of n- or p-type conductivity, respectively. This form of silicon preserves short-range order among the atoms, so that the unit cell structure is the same for both crystalline and amorphous states, and broken bonds that ensure the absence of long-range order are partly restored by elements with positive electron affinity and suitable atomic radius, for example, hydrogen. Further modifications of the optical and electrophysical properties of this material, and improvements in its stability by fluorine or carbon doping, are being investigated [196].

An important advantage of amorphous silicon is that its absorption coefficient α is greater than the absorption coefficient of single-crystal silicon by more than an order of magnitude [197,198]. Figure 2.27 shows the function $\alpha(\lambda)$ for amorphous silicon [198]. Practically all the solar radiation that is photoactive in this material is absorbed within 1.5-2.0 μm, which means that solar cells can be fabricated from a material that is cheaper by a factor of 50-100 as compared with crystalline semiconducting material.

However, studies of the properties of amorphous silicon have shown that it is difficult to exploit it as a material for high-efficiency solar cells because of the short carrier lifetime and short carrier diffusion length ($L = 0.05$-0.1 μm). The band gap of films of amorphous silicon is 1.6-1.8 eV, depending on deposition conditions [197,198]. The highest efficiencies (3-7%) have been attained by using p—i—n structures and Schottky barriers with platinum and chromium. The expansion of the space-charge region in cells of this type ensures that a large proportion of the incident solar radiation

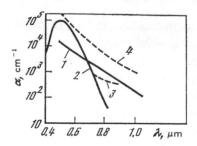

FIG. 2.27. Wavelength dependence of the absorption coefficient of silicon: 1) single crystal; 2) undoped amorphous silicon containing hydrogen; 3, 4) n- and p-type amorphous silicon, respectively.

is absorbed directly in this region so that further increase in efficiency can be assured. Doping with boron or phosphorus produces a higher absorption coefficient (see Fig. 2.27), but reduces the carrier lifetime. This means that the amorphous-silicon cell with the p—n or p—i—n structure has a low collection coefficient in the long-wave part of the spectrum, and a reduced absorption coefficient at shorter wavelengths, when the n^+ and p^+ layer is of poor quality [197]. Structures with Schottky barriers are therefore preferable in this respect because they produce more complete collection of charges created by light in thin surface layers [199].

The open-circuit voltage of these cells is up to 0.8 V, but the generated photocurrent density does not exceed 12 mA/cm^2 for an efficiency of about 5.5% under AM1 illumination [199].

Another unsolved problem in the development of amorphous-silicon thin-film solar cells concerns the reduction in the resistance of the junction between the contact and the semiconducting layer which, in many cells, amounts to 3-10 $\Omega \cdot$cm^2 and produces a deterioration in the current-voltage characteristic and low fill factor.

The properties of thin-film solar cells made from amorphous silicon can be substantially improved by using drift fields, by increasing the conductivity of the p^+ layer and simultaneously increasing its transparency (so that a large proportion of the incident light reaches the space-charge layer), by exploiting multiple reflection of light from boundaries in the interior of the film, and by using a reflecting contact of aluminum, silver or chromium. For example, it is possible to produce $U_{oc} \sim 1.1$ V and I_{sc} of the order of 15-20

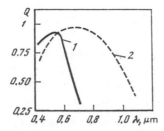

FIG. 2.28. Wavelength dependence of the collection coefficient of silicon solar cells: 1) amorphous silicon with the p—i—n structure and an ITO film facing the incident radiation; 2) single-crystal silicon with the p—n junction at a depth of 0.3 μm.

mA/cm^2 which, even for a fill factor of not more than 0.6, will ensure an efficiency of about 10%. Higher efficiency of amorphous-silicon cells must be combined with improved stability of cell characteristics, since the photoconductivity of low-grade α-Si:H films may fall by a factor of ten or more over the seven or eight hours of continuous illumination, while heating above 300°C results in the onset of exodiffusion of hydrogen from the film, which produces a rapid deterioration in its parameters. The stability and quality of amorphous silicon films are improved by using the three-stage method described in [200].

The first stage is to coat the substrate with the α-Si film without hydrogen (by evaporation in a high vacuum, using an electron beam or the thermal method). The rate of condensation is 2-5 Å//sec. The film is then annealed, which results in a consolidation of the film and an associated reduction in the number and volume of micropores. The α-Si film is then impregnated with hydrogen in a high-current plasma source capable of generating hydrogen ions of 20-25 keV. This saturates the amorphous silicon film with hydrogen down to a depth of 0.3 μm. The three-stage method produces stable, high-quality amorphous silicon films which are probably almost completely free from micropores.

The properties of amorphous silicon films can be stabilized and their photoconductivity increased by laser annealing [201,202], ion implantation [203], and heating the substrate to 200-400°C during the film deposition process [204]. In the case of cells with the p—i—n structure and a transparent conducting tin dioxide window, the efficiency reaches 7.5% and is expected to rise to 10% [205].

Despite the relatively low efficiency, small and economical solar batteries consisting of eight series-connected amorphous silicon cells generating only 4.5 μW/cm^2 under 300 lux are widely used as power sources for small electronic clocks and calculators with liquid crystal displays [198]. The spectral sensitivity of amorphous silicon cells near the ultraviolet region exceeds the sensitivity of solar cells made from single-crystal silicon [198] (Fig. 2.28) and resembles the wavelength dependence of the sensitivity of the human eye, so that such cells have promising applications as exposure meters in photography.

2.8   Lowering the cost and automating the fabrication of solar cells

The expected extensive use of solar cells in the immediate future, not only on board spacecraft, but also in terrestrial solar power engineering, in industry and agriculture, in automatic control systems, and in consumer electronics, means that the development of completely automated fabrication of solar cells from cheap, thin semiconducting layers has become an urgent problem.

The situation is complicated by the fact that, for a long time, attempts to achieve maximum efficiencies and optimum optical and electrical parameters were made without sufficient attention being paid to the thickness of solar cells, and without trying to reduce the cost, or to mechanize and automate the fabrication of solar cells and their assembly into batteries. A great variety of physical and chemical processes and operations has been used. These include the following:

   — in the fabrication of silicon solar cells, the p—n junction was produced by high-temperature thermodiffusion of impurities, preceded by chemical cleaning of the surface with liquid solvents and etching solutions, and the baking-on of contacts in an inert gas atmosphere, followed by the deposition of antireflective coatings by evaporation in high vacuum;
   — the fabrication of gallium arsenide cells was based on expensive, thick substrates and the use of laborious liquid or gas epitaxy to produce layers of the solid solution of aluminum in gallium arsenide;

— the fabrication of thin-film cells was based on the hetero-system copper sulfide—cadmium sulfide and the simultaneous use of "dry" method (deposition of cadmium sulfide layers on conducting substrates) and "wet" method (heterojunction produced by chemical reaction in the liquid phase between the cadmium sulfide surface layer and cuprous monochloride).

It is clear that the problem of automating the fabrication of solar cells will require the development of new manufacturing technologies, including a small number of uniform operations, while the cost of the cells will be reduced by using inexpensive thin layers as well as polymer materials (without sacrificing the optical and electrical properties of the cells).

Substantial changes have already taken place in this area, and the technologies established to produce inexpensive solar cells for terrestrial uses are beginning to be transferred to the fabrication of solar batteries for space applications [206].

The production of silicon by direct reduction of silicon dioxide is being developed. The continuous drawing of silicon strip has been introduced, which means that laborious and expensive operations involving cutting, grinding, and chemical and mechanical polishing can be eliminated. Antireflective coatings, contacts, and films for doping by chemical pulverization have been produced. This technology can be referred to as "chemical." At the same time, "physical" technology has been developed for the fabrication of solar cells in which the antireflective coatings, contacts, and the introduction of the dopant are achieved by ion implantation in a vacuum, and defects produced in the doped layer are annealed by electron- or laser-beam scanning [207]. These operations can follow directly one after another.

As a rule, the new technological processes have been developed for silicon solar cells, but substantial advances have been made in improving the quality and reducing the cost of other cells, too.

For example, thin (10-15 μm) layers of the expensive gallium arsenide have been produced by peeling them off a substrate [208].

When gallium arsenide films are grown epitaxially on a thick single-crystal substrate, the latter is first coated with a solid solution of aluminum and gallium arsenide (4-5 μm), which is then rapidly dissolved in hydrofluoric acid (gallium arsenide is not), so that the thin film of gallium arsenide can readily be detached from the substrate. The difference between the melting points of individual layers can be used to detach thin films or sets of multilayer epitaxial heterostructures.

The cost of solar cells based on homojunctions and heterojunctions in gallium arsenide can be reduced and their characteristics improved by producing these structures on the relatively cheap germanium substrates [208] and by developing two-sided cells with p—n junctions on the front and back surfaces of the gallium arsenide [209]. The design of these cells is similar to that developed for silicon solar cells (see Section 2.5). Thin-film gallium arsenide cells with a shallow p—n junction in a homogeneous medium can be fabricated by using gallium arsenide films produced by gas epitaxy on silicon substrates coated with a thin layer of germanium [210]. Such cells with the n^+—p—p^+ structure have been found to have efficiencies in excess of 12% under AM1 illumination.

Unfortunately, numerous attempts to use natural and synthetic organic materials for solar cells have so far been unsuccessful. For example, for a solar cell containing a thin layer (0.1-0.3 μm) of the dye, magnesium phthalocyanine (the contacts were copper, gold, or silver layers and the separating barrier was in the form of a semi-transparent aluminum layer), the photoionization quantum yield was found to be 33% at λ = 0.65 μm, and there was appreciable spectral sensitivity in the wavelength range 0.5-0.8 μm. However, the efficiency of these cells was well below 1% [211]. A solar cell of similar composition with a 0.8—1.5 μm photosensitive layer (consisting of phthalocyanine particles dispersed in a polymer film) and an indium contact can generate I_{sc} = 0.8 μA/cm^2 and U_{oc} = = 0.34 V. Its efficiency under terrestrial conditions was of the order of 1%. When the concentration ratio was increased by a factor of 14, the current I_{sc} was found to rise to 1.5 mA/cm^2 and U_{oc} to 0.5 V. The cell was assembled on a glass substrate coated with a transparent conducting film.

It is perfectly possible to use the natural pigment, chlorophyll, as the active photosensitive material in organic solar cells [213]. The optical and electrical properties of electrochemical solar cells whose efficiencies at present do not exceed 9%, and whose photosensitive semiconducting electrodes are separated by a layer of an electrolyte, have been under investigation by numerous researchers. A reasonably detailed review of this work can be found in [214].

Despite the substantial advances in reducing costs and in simplifying the fabrication technology, the cell-manufacturing process still consists of tens of laborious operations. It is likely that only a radical change in the basic production stages will allow us to automate fully their manufacture in the near future. This difficulty may be resolved in practice by using new types of solar cell that are being actively developed at present. The optical and photoelectric parameters of these cells are somewhat different from those of traditional cells with the p—n junction in homogeneous and heterogeneous semiconducting materials (in particular, they have higher sensitivity in the ultraviolet).

Most of these new cells are modifications of the Schottky barrier, i.e., the barrier between the semitransparent metal layer and the semiconductor [215,216]. The rapid band bending near semiconductor metal boundaries produces the separating barrier necessary for the operation of the solar cell. The semitransparent metal can be replaced by semitransparent conducting oxides of wide-band semiconducting materials [217,218].

The parameters of these solar cells have gradually improved and, through the optimization of the properties of the semitransparent metal films, they have now reached the level typical for single-crystal silicon solar cells incorporating the p—n junction [219]. The expensive metal layers (silver, chromium, or gold) used in the case of the silicon base have been successfully replaced with semitransparent layers of the very much cheaper aluminum [220]. The open-circuit voltage has been substantially increased, and the saturation reverse current reduced, by introducing a very thin (10-20 Å) layer of silicon dioxide on the boundary of the Schottky barrier [221,222], grown on the barrier before the semiconducting semitransparent layer is deposited on the surface of the silicon. This silicon dioxide film can be produced by thermal oxidation in air or in oxygen at 400°C for 20-30 min, by electrochemical anodic oxidation, or by immersion in a solution of hydrogen peroxide for 2-15 min at 60-70°C [223].

A high positive charge is usually present in the intermediate oxide layer, but subsequent deposition of the conducting tin dioxide film, containing hydrogen and chlorine ions, compensates this charge, so that the technique could be used to produce cells with η = 14% (I_{sc} = 36 mA/cm^2, U_{oc} = 0.525 V, fill factor 0.74) on n-type single-crystal silicon [224]. The band bending on the surface of the semiconductor and the separating barrier can also be produced by having an oxide layer with a strong built-in charge [225]. Like the diffused doped layer, this inversion layer can also be used to reduce the rate of surface recombination on the illuminated surface. The separation of carriers is then performed on the p$^+$ and n$^+$ barriers on the base layer located at the back of the cell [226,227]. This facilitates the assembly of cells into the groups and modules used in solar batteries by means of printing techniques.

The fabrication of the above types of solar cell compares favorably with high-temperature thermodiffusion (800-900°C) used to produce p—n junctions in silicon because the separating barrier can be introduced at a relatively low temperature (200-400°C). Schottky barriers and MOS or SOS structures (metal—oxide—semiconductor or semiconductor—oxide—semiconductor) can be produced not only on single-crystal materials, but also on polycrystalline and strip media, as well as on amorphous silicon films. The deposition of Schottky barriers or MOS structures can be performed in the same technological cycle in which the contacts are introduced and anti-reflective coatings are deposited. There are two ways of doing this:

— all operations can be performed in the same vacuum chamber or in a number of chambers connected through flap valves, using masking templates or "dry" photolithography;

— all operations can be performed in air or in an inert gas by chemical pulverization, silk screening, and chemical or electrochemical deposition.

However, it is important to note that the high efficiencies (12-15% under terrestrial conditions) of solar cells incorporating Schottky barriers and MOS and SOS structures are usually produced by employing isotype p—p$^+$ or n—n$^+$ barriers at the rear of the base layers or doped regions under contacts formed by thermodiffusion of impurities [223-227]. The doping operation is, of course, a further stage that is not readily automated. The isotype barrier in the base of solar cells incorporating the Schottky barrier and MOS or SOS structures was produced by fusing aluminum that was first deposited on the back surface by thermal evaporation in a vacuum, or was applied in air in the form of a paste with an organic binder [220,

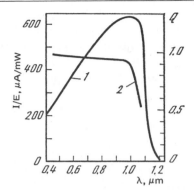

FIG. 2.29. Wavelength dependence of the sensitivity (1) and collection coefficient (2) of a solar cell with the following structure: ITO SiO_x film-single crystal n-type silicon with resistivity $\rho = 10\ \Omega \cdot cm$.

228] . MOS and SOS structures, similar to those used to prepare the top separating barrier, were also introduced [229,230] . In particular, $Al-SiO_x-p-Si$ layers were used to produce a separating barrier while $Pt-SiO_x-p-Si$ layers were used for the isotype back barrier (the platinum layer can be continuous) [227] . These improvements suggest that isotype barriers can be produced in the solar-cell base by a single automated technological cycle.

The wavelength dependence of the sensitivity and collection coefficient of solar cells, made from single-crystal silicon with the SOS structure and a thin intermediate layer (a few tens of Ångstroms) of SiO_x on the silicon surface, is shown in Fig. 2.29 [223] . The top contacting transparent layer (ITO film) was deposited by chemical pulverization from a mixture of indium and tin oxides. The thickness of this layer was 700 Å (surface resistance of about 120 $\Omega/\square$), so that the film also acted as an efficient antireflective coating. The efficiency of these cells under AM0 illumination was found to be 10.8%. This figure can be substantially increased by reducing the series resistance and, in particular, by optimizing the parameters of the ITO film as well as the dimensions and thickness of the contact grid on the illuminated cell surface.

Stability tests [213] performed on the above solar cells must exhibit the physicochemical compatibility of the thin intermediate layer and the barrier metal or oxide under continuous illumination and increased temperature. However, there is no doubt that careful hermetization and isolation from the ambient medium will be necessary to ensure prolonged operation of these new solar cells.

2.9 The future of solar cells with maximum efficiency and maximum collection coefficient throughout the solar spectrum

The publication of the first papers reporting maximum attainable solar cell efficiencies of 24-25% [15,67,81,82,85] was probably followed by searches for new physical ideas that could be exploited to produce more efficient solar cells. The cascade and multijunction cells, and cells with heterojunctions, built-in electric fields, and graded structures [13,64,80,84,116], which appeared soon after, could not be subjected to experimental tests for a long period of time, although the limiting efficiency of most of these new cells was as high as 30-50%. The successful practical implementation of many of these new ideas [5,77,117] resulted in AMO efficiencies of 17% in the case of the cheap silicon cells with base layer produced by the casting method [232], and efficiencies of 20-25% in the case of cells incorporating heterostructures in a system consisting of the solid solution of aluminum in gallium arsenide and gallium arsenide [161-164,166-168].

Theoretical investigations directed toward the development and improvement of the cell based on the bulk photoelectric effect in semiconducting structures and intended for solar energy conversion, showed that the complete absorption and conversion of solar photons into electric power was possible in such cells if they consisted of a graded-gap structure in which the maximum band gap (on the surface) corresponded to the short-wave limit of the solar spectrum, and the minimum gap (on the separating barrier near the back surface) corresponded to the long-wave limit, while the ratio of electron-to-hole mobilities was high. This virtually removed the restriction on the maximum efficiency of conversion of solar optical radiation by semiconducting solar cells, and efficiencies of 90% were indicated as possible [233-235].

Several ways were suggested for attaining high efficiencies of conversion of solar radiation directly into electric power [2,5,234].

Theory suggests (and this is confirmed experimentally) that the efficiency of a solar cell increases with increasing irradiance

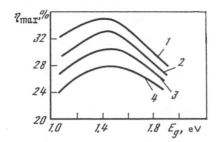

FIG. 2.30. Maximum efficiency of a solar cell containing the p—n junction in a homogeneous material as a function of the gap width of the semiconductor for different concentration factors: 1) 1000; 2) 100; 3) 10; 4) 1 (direct solar flux).

[19,164,166,234,236]. The Dember emf, which appears within the body of the semiconductor, is due to the difference between the mobilities of electrons and holes created by sunlight and their interaction. When the incident light falls on the p^+ layer in the solar cell with the p^+—p—n^+ structure, the Dember emf has the same sign as the photo-emf generated by the isotype barrier, but the signs are different when the top n^+—p barrier is illuminated. This means that, to produce higher values of U_{oc}, it is useful to illuminate the two-sided cell with higher intensity at the back (in high light fluxes). The number of excess carriers produced under high irradiance is much greater than the number in thermal equilibrium (determined by dopant concentration in the semiconductor), which leads to a reduction in the resistivity of the base as a result of the phenomenon of photoconductivity. Under very high irradiance (higher by a factor of the order of 1000), the value of U_{oc} that can be produced in these solar cells is close to the band-gap potential of the semiconductor, expressed in volts. The higher efficiency of solar cells under high-irradiance AM1 illumination can be represented by the calculated dependence of maximum efficiency on the band gap of the semiconductor at 300 K (Fig. 2.30) [236]. A similar function was calculated in [82] under onefold irradiance.

The efficiency of the solar cell consisting of the optimal semiconducting material with a band gap of about 1.4 eV can be raised to 35% but only by increasing the solar flux density by a factor of 1000 (Fig. 2.30, curve 1).

These calculations have stimulated the development of terrestrial photogenerators containing solar cells operating under very high degrees of concentration of solar radiation, for example, up to 440 in [237] with the likelihood of increasing this figure to 2000-2200.

This method of increasing the solar cell efficiency already requires the resolution of some engineering and fabrication difficulties due to the necessity for removing large quantities of excess heat in order to maintain the cells at a low enough temperature. Moreover, since the advent of solar concentrators with stable coatings and long life [23], two other promising ways of increasing the solar conversion efficiency by a substantial factor have been under laboratory investigation. On the face of it, these two methods are diametrically opposite.

The first requires a drastic compression of the wide-band solar spectrum and the subsequent transformation of this radiation into electric power by a solar cell containing the p—n junction in a homogeneous semiconducting material whose band gap is matched exactly to the spectrum of solar radiation intercepted by it. It is well known that, when the spectrum is compressed, the solar-cell efficiency becomes much greater because losses through nonphotoactive absorption at long wavelengths are no longer significant and there are no thermal losses through thermal scattering of short-wave photons.

The second method relies on the cascade arrangement of several solar cells that are transparent at long wavelengths, beyond the fundamental absorption edge, and each cell transforms efficiently the corresponding portion of the incident radiation, so that this system covers the entire solar spectrum and effectively extends the spectral sensitivity of the solar cell.

Natural solar radiation can be compressed in a variety of ways. For example, the concentrated beam of solar radiation can be directed onto a receiver in the form of a black body with a thermally stable selective radiator coated with erbium oxide which radiates preferentially between 1 and 2 μm [239]. Germanium [239] or silicon [240] solar cells convert the spectrally compressed radiation (practically without loss) with an efficiency in excess of 25%. In the case of cells with a greater band gap, for example, cells made of gallium arsenide with a homojunction or heterojunction, it will be

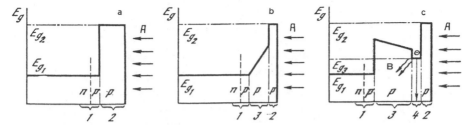

FIG. 2.31. Band diagrams and disposition of layers in different solar cells based on gallium arsenide: a) junction in a homogeneous material and wide-gap filter-window on the surface; b) graded-gap structure in the filter-window; c) reradiating structure between the two regions of the filter-window; 1) p–n junction in gallium arsenide; 2) filter-window consisting of a solid solution of aluminum in gallium arsenide; 3) graded-gap structure (variable-x composition of $Al_xGa_{1-x}As$); 4) reradiating structure; A) solar radiation; B) emission.

necessary to develop a selective thermal radiator working at shorter wavelengths, which will result in still higher practical efficiencies.

The solar spectrum can also be compressed by semiconducting light diodes incorporating heterostructures in gallium arsenide, which convert short-wave radiation into long-wave radiation with almost 100% quantum efficiency. The long-wave radiation is matched in energy to the band gap of homogeneous gallium arsenide [115,241]. It has also been suggested that a monolithic multilayer solar cell might incorporate this type of reradiating structure with a solar converter based on a combination of the solid solution of aluminum in gallium arsenide and gallium arsenide [117,166].

Figure 2.31 shows the band diagrams of different solar cells based on gallium arsenide. In the case of a reradiating structure between the two regions of the filter-window, the region facing the incident light (region 2) has the composition $Al_{0.8}Ga_{0.2}As$, and the reradiating structure (region 4) has the composition $Al_{0.1}Ga_{0.9}As$ with the aluminum concentration gradually increasing (region 3) to 0.3 as the p–n junction in gallium arsenide is approached (region 1). In this structure, the photosensitivity spectrum is determined by the thin (less than 1 μm) uppermost region 2 of the wide-gap filter-window, and the spreading resistance is reduced by the relatively thick (20-30 μm) inner region 3 of the filter-window which is transparent to the long-wave luminescence emitted by the reradiating

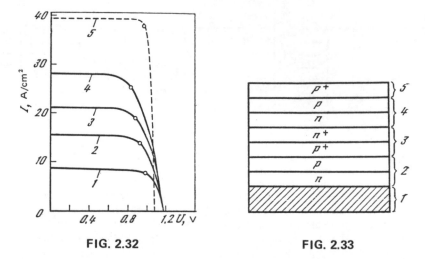

FIG. 2.32   FIG. 2.33

FIG. 2.32. Current-voltage load characteristics of solar cells based on gallium arsenide with a heterojunction and reradiating structure for different concentrations of terrestrial solar radiation: 1) 530; 2) 970; 3) 1350; 4) 1800; 5) 2570 (calculated from measurements obtained when an 0.9-mm diameter spot was illuminated on a 1-cm diameter cell).

FIG. 2.33. Disposition of layers in a monolithic cascade structure: 1) n-type substrate of single-crystal gallium arsenide; 2) solar cell of gallium arsenide with a p—n junction in a homogeneous material; 3) tunnel junction made from highly doped solid solution AlGaAs; 4) solar cell incorporating an AlGaAs—GaAs heterojunction and a p—n junction in gallium arsenide; 5) wide-gap filter-window.

structure 4 toward the p—n junction in the gallium arsenide (region 1) after absorption of the solar radiation [117,166].

Solar cells with a reradiating structure between the filter-window regions are particularly suitable for the transformation of low-intensity solar radiation because they have a wide spectral sensitivity range and low series resistance. The current-voltage characteristics of cells based on gallium arsenide with a hetero-junction and a reradiating structure are shown in Fig. 2.32 for different concentration ratios (up to 2570) [242]. The maximum electric power developed across the load of one such cell, 1 cm in diameter, was shown by terrestrial measurements to be 13.5 W. Consequently, eight such solar cells with concentrators [117,166,242] are necessary to produce more than 100 W of electric power, while

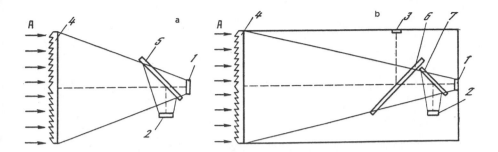

FIG. 2.34. Cascade system incorporating two solar cells and one dichroic mirror (a) or three solar cells and two dichroic mirrors (b): A) sunlight; 1-3) solar cells; 4) Fresnel lens; 5-7) dichroic mirrors.

the same electric power can be produced under terrestrial conditions by conventional high-grade solar cells in the form of a flat panel of area not less than 1 m² (more than 10,000 solar cells of area 1 cm² each).

It is clear that the cost of the laborious manufacturing technology used to produce the new solar cells with complex multilayer structure is completely recovered, and it is possible that, when such cells are widely used, the cost of electric power generated by solar cells will be reduced by two or three orders, so that it will approach the cost of power generation by traditional sources (thermal and hydroelectric power stations).

The development of the highly efficient homo- and heterostructures on silicon and gallium arsenide has led to increasing interest in cascade solar cells consisting of these structures. This has given rise to particular technological and fabrication difficulties [108], and attempts have therefore been made to develop cascade cells with a single monolithic structure, produced by successive deposition of layers on a gallium arsenide substrate by liquid, gas, or molecular epitaxy, as shown in Fig. 2.33 [243]. The top (2) and bottom (4) solar cells in this two-cascade system are joined in series by the n^+-p^+ tunnel junction of AlGaAs (Fig. 2.33, region 3). The prototype structure [243] produced U_{oc} of about 2.2 V, but the relatively low current and its efficiency did not exceed 10-15%, probably because of the high resistance of the tunnel junction (0.58 Ω).

The high quality of the tunnel junctions reported in [244]

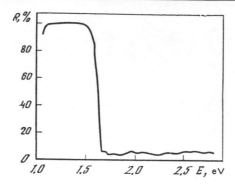

FIG. 2.35. Reflection coefficient of an efficient 17-layer dichroic mirror for cascade solar cells as a function of energy.

offers us the hope of further advances in the development of mono-lithic cascade solar cells.

Much greater success has been achieved by researchers using two or three solar cells arranged at right angles to one another. The incident solar radiation is concentrated by a Fresnel lens and falls on one or two multilayer dichroic mirrors which split the spectrum into individual segments, directing onto each cell the particular spectral interval in which it has maximum sensitivity (Fig. 2.34) [245,246].

In practical implementations of these ideas, the efficiencies of the individual cells (they must be different in different spectral ranges) and the high quality and stability of the dichroic mirror (under prolonged continuous operation) are of great importance.

The wavelength dependence of the reflection coefficient of the dichroic mirror produced by the successive deposition in vacuum of 17 transparent films of ZnS (n = 2.3) and Na_3AlF_6 (n = 1.35) is shown in Fig. 2.35 [114]. The radiation transmitted by the mirror was received by a solar cell based on gallium arsenide, whereas the reflected radiation was received by a silicon cell (see Fig. 2.34a).

The two-stage system with the dichroic mirror [114] and the 165-fold concentration of terrestrial solar radiation (flux density 894 W/m^2, incident spectrum corresponding to AM 1.23) has the following cell parameters measured at 30°C (water cooling):

	E_g, eV	I_{sc}, A	U_{oc}, V	F^*	η, %
AlGaAs	1.61	1.382	1.26	0.827	17.4
Si	1.1	1.711	0.738	0.725	11.1

*F is the fill factor of the current-voltage characteristic of the solar cell.

The resultant efficiency of the two-stage system is 28.5%.

The improved quality of dichroic mirrors and of the individual solar cells has meant that efficiencies of 30-32% have been achieved in these split-spectrum systems for a concentration factor of 50-100, while an efficiency of 40% has been attained for higher concentration ratios (of the order of 1000) [114,243-246].

It is best to use the following semiconducting materials in cascade system with dichroic mirrors [114]: Ge for $E_g = 0.7$ eV; Si, In_xGa_{1-x}, $GaAl_{1-x}Sb_x$, $Ga_yIn_{1-y}As_{1-x}P_x$, $Al_yGa_{1-y}As_{1-x}Sb_x$ for $E_g = 1.1$ eV; GaAs for $E_g = 1.4$ eV; and $Al_xGa_{1-x}As$, $GaAs_{1-x}\cdot P_x$, $Al_yGa_{1-y}As_{1-x}Sb_x$ for $E_g = 1.7$ eV.

Calculations have shown that the cascade solar cell of this type, which consists of an AlGaAsSb ($E_g = 1.8$) and a GaAsSb ($E_g = 1.2$ eV) cell, should have an efficiency of 29% at 350 K when the concentration ratio is 100 [60].

It is important to note that systems incorporating dichroic mirrors avoid the basic problem of combining, in the cascade cell, a number of layers with similar lattice constants and thermal expansion coefficients (Fig. 2.33). This problem arises when monolithic cascade cells are produced by epitaxial growing of layers.

It is possible that there will be no need for dichroic mirrors in the future: solar cells (see Section 2.4) that are transparent in the longwave part of the spectrum (beyond the fundamental absorption edge) and have a highly reflecting metal coating or mirror at the rear will be used to convert solar radiation and split the solar spectrum (Section 2.4).

2.10 Optical and electron-microscope methods of investigating the properties of solar cells

Studies of the properties of solar cells include measurements of the spectral sensitivity, the collection coefficient, and the current-voltage load characteristic (as well as the dark characteristic for different voltages applied in the forward and reverse directions

[15-22, 69-71]). They also involve measurements of the spectral distribution of the coefficients of reflection, transmission, and absorption (mostly for selective optical coatings [23]), and the X-ray and physicochemical analysis of the composition and structure of semiconducting materials and layers [247,248]. Non-traditional optical and electron-microscope methods of studying the properties of solar cells and their individual layers are of particular interest in the investigation of cells consisting of different single-crystal and thin-film semiconducting materials.

Parallel studies of photoluminescence spectra
and optical and structural properties

Photoconductivity and long-wave photoluminescence produced when semiconducting crystals and layers are illuminated by short-wave optical radiation are usefully investigated at the same time by measuring the optical and structural properties of the various materials. The photoluminescence and photoconductivity spectra of semiconductors can be interpreted particularly fully. This composite analysis of the properties of solar cells based on gallium arsenide with a p–n junction in the homogeneous semiconducting material, and a heterojunction on the outer surface in the form of a wide-gap filter-window of $Al_xGa_{1-x}As$, was carried out in [249] for different aluminum concentrations. The highest grade solar cells investigated had two photoluminescence bands, namely, a strong narrow band with an emission maximum at 1.48-1.49 eV, due to edge emission by p-GaAs, and the fundamental emission band of the wide-gap filter-window [249]. The position of this band along the wavelength axis was found to depend on the amount of aluminum in the solid solution. For example, when x = 0.4, the maximum of the fundamental emission band of the filter-window was at 1.95 eV. No other emission bands were found in the photoluminescence spectra of high-efficiency solar cells because the impurities were gettered, while the $Al_xGa_{1-x}As$ layer was grown by liquid epitaxy in the presence of excess gallium and aluminum, which facilitated the attainment of a longer carrier lifetime in the space-charge region, a reduction in the recombinational component of the dark current down to 10^{-10} A/cm^2, and the appearance of a second exponential segment on the forward branch of the dark current-voltage characteristic with $I_0 = 10^{-13}$ A/cm^2 and A = 1.5–1.6.

According to Golovner et al. [249], the two peaks at 1.33 and 1.4 eV in the photoluminescence spectra of individual solar cells are probably due to carrier transitions to the ground acceptor level of copper (E = 0.15 eV), or a lower lying acceptor level of silicon or tin (E = 0.08 eV). These transitions are due to the appearance of additional recombination channels, which leads to a deterioration in the properties of the p–n junctions and in the photoelectric parameters of solar cells generally. The photoluminescence spectra were recorded simultaneously with the reflection spectra of the surface of solar cells with different concentrations of aluminum in the wide-gap filter-window (using the reflection attachment to the Hitachi-225 infrared spectrophotometer).

The regions of enhanced reflection in these spectra correspond to lattice absorption bands of the original gallium arsenide (wavelength interval 35-40 μm) and of aluminum arsenide (AlAs, 28-30 μm). The reflection coefficient minimum of the surface of solar cells with a wide-gap filter-window, which occurs at wavelengths of 30-35 μm, shifts toward longer wavelengths as the percentage concentration of aluminum is reduced. The intensity of the maximum is reduced at the same time, and the reflection bands become narrower [249]. It is thus clear that optical measurements can be used to determine the composition of the wide-gap filter-window of solar cells (especially when its thickness is sufficient, i.e., between 5 and 30 μm). An X-ray microanalyzer [249] was used in control measurements (in combination with optical measurements) and gave the band gap of the solid solution as a function of its composition, which turned out to be similar to that reported in [165].

The photoluminescence spectra excited in thin-film front-barrier solar cells based on the copper sulfide–cadmium sulfide heterosystem were investigated with the aid of the nitrogen laser ($\lambda = 0.3371$ μm) and the DKSSh-1000 lamp working in conjunction with a filter defining a narrow band near $\lambda = 0.365$ μm [250]. The photoluminescence was collected by an elliptic mirror with a flexible optical fiber at its focus. The fiber transferred radiation from the specimen chamber to the entrance slit of the MDR-2 monochromator (the cathodoluminescence spectra were recorded in a similar way).

Practically all the photoluminescence bands in the characteristic

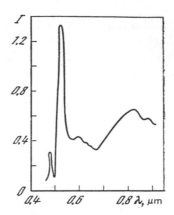

FIG. 2.36. Photoluminescence spectrum of thin-film front-barrier photocells based on the copper sulfide—cadmium sulfide heterosystem.

photoluminescence spectrum of thin-film solar cells based on the copper sulfide—cadmium sulfide heterosystem (Fig. 2.36) are due to cadmium sulfide (they include the excitonic emission at 0.492 μm, the orange emission at 0.615 μm, due to deep acceptor centers formed by copper and sulfur vacancies, and the band at 0.52 and 0.83 μm). It is probable that this is so because of the small thickness of the copper sulfide layer on protrusions consisting of cadmium sulfide crystallites. Only some of the spectra showed the presence of the 0.96-μm chalcocite peak. However, this band lay in the region of low sensitivity of the photomultiplier used to record the radiation transmitted by the MDR-2 monochromator. The photoluminescence spectra of this layer could be recorded more clearly as functions of its stoichiometric composition by improving the method of measurement and by using thin-film solar cells with a sufficiently thick layer of copper sulfide, prepared specially for these studies.

Composition and properties of semiconducting layers, determined from ultraviolet and infrared reflection data

The depth of surface defects produced when silicon is subjected to mechanical treatment can be determined from measurements on the ultraviolet reflection peak at 0.28 μm, which is characteristic

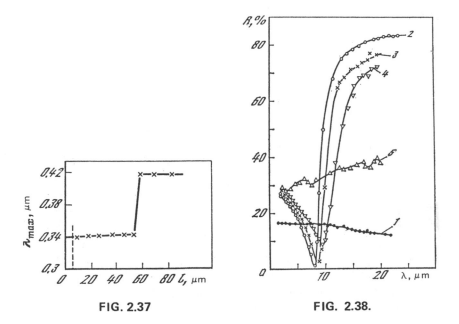

FIG. 2.37.   FIG. 2.38.

FIG. 2.37. Position of the ultraviolet reflection peak as a function of the thickness of the gallium phosphide layer removed by mechanical grinding from the surface of the GaP—GaAs heterojunction: dashed line) p—n junction in gallium phosphide, produced by thermodiffusion of zinc.

FIG. 2.38. Infrared reflection coefficient of the n-GaAs (150 μm) —n-Ge heterojunction as a function of wavelength: 1) initial state; 2) after the removal of surface layers of gallium arsenide down to the boundary with the highly doped germanium for an electron concentration of $3 \cdot 10^{19}$ cm^{-3}; 3-5) after the removal of a layer of germanium, 3-5, 8-12, and 15 μm thick, respectively.

for this material. The appreciable difference between the positions of the ultraviolet reflection peaks of most common semiconducting materials [25,185], which depends on their band structure [10,11, 58], means that it is possible to use optical methods to determine with high precision the thickness and composition of the individual layers in heterostructures.

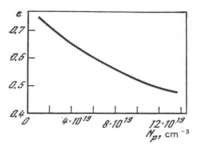

FIG. 2.39. Integrated thermal emission coefficient at 30°C as a function of the concentration of free carriers in p-type gallium arsenide.

The properties of solar cells and heterojunctions based on the GaP—GaAs heterostructure have been investigated on relatively thick gallium phosphide layers produced by diffusing phosphorus into gallium arsenide [251]. The reflection coefficient was carefully measured in the ultraviolet and visible ranges of the spectrum as 5-6 μm layers were carefully removed by mechanical grinding of the gallium-phosphide surface. At least 60 μm of gallium phosphide had to be removed before the 0.34-μm ultraviolet reflection peak was found to shift to 0.42 μm (Fig. 2.37), the position typical for single-crystal gallium arsenide. The position of the ultraviolet peak does not depend on the impurity concentration or type of conductivity. This means that it is possible to determine accurately the thickness of the entire layer of gallium phosphide and determine the position of the heterojunction, which is also facilitated by the high ultraviolet absorption coefficient of both layers (control measurements reveal that the gallium phosphide layers become transparent for $\lambda \geqslant 0.5 \, \mu$m).

When the semiconducting layers in heterojunctions have sufficiently different dopant concentrations, the boundary between them can readily be determined from the infrared reflection spectra. This was used to investigate the properties of gallium arsenide—germanium heterojunctions, where the 150-μm epitaxially grown gallium arsenide layer was produced on thick n- or p-type germanium substrates. It was assumed that the surface layer of germanium on the boundary with the gallium arsenide film was enriched with arsenic in the course of epitaxy, and was converted into a layer of degenerate n-type germanium [252]. Optical measurements of the infrared reflection coefficient of these heterostructures, as successive layers were removed from the gallium arsenide side (Fig. 2.38), not only confirmed this assumption, but also yielded the thickness of the highly doped germanium layer. The smooth monotonic wavelength dependence of the reflection coefficient of the surface of the gallium arsenide layer (curve 1, Fig. 2.38) was replaced after this layer was removed by a curve with a clearly defined plasma resonance minimum (curve 2) which is characteristic for highly doped semiconductors. The dopant concentration in this layer was found to fall with distance from the heterojunction boundary (curves 3 and 4) until (after about 15 μm of the layer had been removed) the reflection coefficient was practically completely determined by the relatively weakly doped germanium substrate (curve 5). The presence of arsenic in germanium near the heterojunction boundary was confirmed by X-ray microanalysis [252]. The electron beam incident on the specimen was scanned in steps of 5 μm over an oblique cut formed at an angle of $3°$. The k_α line of arsenic was recorded for an accelerating voltage of 50 kV. The distribution of arsenic concentration obtained in the course of the X-ray analysis was in qualitative agreement with optical measurements in the infrared.

The well-defined dependence of the infrared reflection coefficient on free-carrier concentration in semiconducting crystals and layers can be exploited in optical studies by employing spectral measurements which are not very laborious but are relatively long. More rapid methods rely on the determination of the integrated thermal emission coefficient of semiconducting specimens, for example, with the aid of the FM-63 thermoradiometer. The latter method requires only a few seconds [23]. The integrated thermal emissivity of opaque specimens is entirely determined by the wavelength dependence of the infrared reflection coefficient. The dependence of the thermal emission coefficient ϵ on the free-carrier concentration of p-type gallium arsenide is shown in Fig. 2.39 [42]. Hall-effect measurements were made to check these values of carrier concentration.

Effect of light, temperature, and corpuscular radiation
on the spectral sensitivity and efficiency of solar cells

The effect of different external factors on the optical and electrical properties of solar cells is usually examined in order to determine the life of such cells under operating conditions and to establish whether the characteristics of cells can be modified in some required manner. These two objectives can sometimes be achieved simultaneously. For example, detailed studies of the effect of corpuscular radiation on solar cells and their selective optical coatings [13, 20-23,96] can now be used not only to predict with sufficient accuracy the output power (and the rate of its deterioration) of solar cells for spacecraft that repeatedly cross the Earth's radiation belts, but also to modify the shape of the spectral sensitivity curve of such cells. Exposure of solar cell batteries to electrons and protons of sufficiently high energy (of the order of a few MeV) leads to a rapid reduction in the carrier diffusion length in the base layer and a deterioration in the long-wave part of the wavelength dependence of the collection coefficient. This means that the spectral dependence of the cell sensitivity can be modified so that, for example, it can be brought closer to the sensitivity of the human eye, with a maximum at 0.5-0.6 μm, without using absorption or interference filters. At the same time, the effect of low-energy protons (30-50 keV), absorbed in the doped top layer of solar cells, reduces the efficiency of collection of carriers from this layer. The spectral sensitivity curve shifts in the reverse direction and, for sufficiently low-energy proton doses, only the long-wave portion of the spectral sensitivity remains in the near infrared. By varying the incident-particle energy, and suitably choosing their flux density, it is possible to produce solar cells that are "blind" to the ultraviolet or even visible solar radiation.

The reduction in and subsequent rise of the spectral sensitivity of silicon solar cells as the incident radiation flux increases are of considerable interest [253,254].

When silicon solar cells are bombarded with 400-keV protons at 45° to the surface, a layer containing defects is formed behind the p—n junction. It has a width of about 1 μm and an enhanced resistivity. The spectral sensitivity deteriorates sharply when the proton flux density reaches 10^{12} cm^{-2}. When this figure is raised to 10^{13} cm^{-2}, the width of the space-charge region is found to extend to the entire damaged high-resistivity region near the p—n junction (on the side of the base layer), which improves the collection of

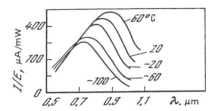

FIG. 2.40. Wavelength dependence of the sensitivity of a silicon solar cell with a p—n junction at different temperatures.

carriers produced by the incident light through the drift mechanism, and the spectral sensitivity increases again.

A reduction in temperature produces an appreciable reduction in the spectral sensitivity of silicon solar cells in the long-wave part of the spectrum (and some increase in the short-wave part; Fig. 2.40) [17]. The temperature dependence of spectral sensitivity is determined by two factors: (1) narrowing of the band gap and (2) the increase in the light absorption coefficient (and, accordingly, a reduction in the depth of penetration into the semiconductor) with increasing temperature.

The temperature dependence of spectral sensitivity is also found to determine the rise in the short-circuit current I_{sc} of solar cells under illumination, but the open-circuit voltage U_{oc} falls more rapidly with increasing temperature than the increase in I_{sc}, which leads to a negative temperature coefficient of efficiency [13,21]. As already noted, a wider band gap of the semiconducting material of the solar cell produces a sharp reduction in the rate of fall of efficiency with increasing temperature (it is lower by a factor of two in gallium arsenide as compared with silicon [19,142].

It had long been considered that the only damage produced by the solar radiation itself was the darkening of the optical coating of the solar cells. However, the development of optically stable multilayer coatings in which the uppermost layer is a glass slide containing cerium dioxide that absorbs the entire ultraviolet radiation below 0.36 μm has reduced cell degradation due to the deterioration in the optical properties of the coating down to a very low level (0.5-2.5%) even under continuous operation on board spacecraft remaining in orbit for several years [23,255].

It was therefore surprising to many researchers to find that the

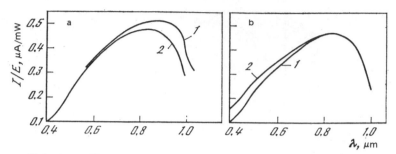

FIG. 2.41. Wavelength dependence of the sensitivity of n^+–p-type (a) and p^+–n-type (b) solar cells, before (1) and after (2) photon degradation.

properties of solar cells deteriorated under exposure to solar radiation (this is the so-called photon degradation). In the initial experiments, this was investigated at the same time as the damaging effect of corpuscular radiation and of temperature [256,257]. These and subsequent studies [258-260] revealed the importance of the simultaneous effect of several damaging factors on the properties of semiconducting materials and of solar cells. These experiments reflect the real practical situation, both in space and in the laboratory. For example, solar cells containing low levels of oxygen in the original silicon wafers produced by zone-melting exhibit high levels of photon degradation, i.e., the reduction in current due to high-intensity illumination may amount to 10-12% [258]. On the other hand, experiments performed without illumination showed that these cells were more radiation-resistant than cells containing silicon grown by the Czochralski method and containing relatively high levels of oxygen. It is possible that the deterioration in the properties of solar cells made from oxygen-free silicon is due to the higher density of dislocations in such crystals. Strong illumination leads to the freeing and activation of boron-containing point defects trapped by dislocations [257]. However, there is no doubt that the additional introduction of oxygen and carbon does stabilize the behavior of solar cells under illumination, especially if the overall concentration of carbon and oxygen in silicon exceeds 10^{17} cm^{-3} [258].

Saturation usually sets in during the photon degradation process after AM0 illumination for 20-40 hr at near-room temperatures, but this time is shorter when the cell temperature rises to 50-60°C [257].

When the n^+—p-type solar cell (the top n^+-layer is produced by phosphorus diffusion) is illuminated or a forward bias is applied to it, the output power is reduced and there is an appreciable reduction in long-wave sensitivity (Fig. 2.41a). On the other hand, the reverse effect is observed for p^+—n-type cells (when the incident spectrum contains wavelengths between 0.35 and 0.45 μm), namely, the output power increases and the spectral sensitivity is reduced in the short-wave region (Fig. 2.41b). The deterioration in the collection of carriers from the base layer of n^+—p-type solar cells is due to the presence of recombination state at 0.37 eV below the conduction band. This level is usually electrically neutral but becomes active under high levels of illumination or because of electrical injection of carriers into the material. The recombination state appears because of the presence in the silicon of a complex consisting of a defect and a silver atom, and, possibly, cluster formations (series of disturbed atoms). Photon degradation can be substantially reduced by preventing silver atoms from reaching the silicon base layer, by mechanically removing the damaged silicon surface layer prior to diffusion, and by diffusing the dopant at a temperature of 875°C or less. For example, when diffusion is performed at 950°C, photon degradation (under illumination by a tungsten halogen lamp producing 1000 W/m^2) amounts to 3-6%. The corresponding figures for diffusion temperatures of 900°C and 875°C are 1-3% and only 0.5%, respectively [259].

The use of a 5-cm water filter that removes long-wave solar radiation or the tungsten-lamp radiation with $\lambda \geqslant 1.2$ μm, will virtually eliminate photon degradation in n^+—p-type cells that is due to the p-type silicon layer [259]. However, the reasons for the improvement in the spectral characteristics of p^+—n-type cells under illumination remain unclear although this phenomenon has been reliably confirmed [259,260] together with the reduction in U_{oc} (under low levels of illumination during measurement), which is particularly well defined when the faces of the cells are strongly illuminated. It is likely that the reduction in U_{oc} in this case is due to losses through defect-containing silicon layers [260].

Photon degradation must be taken into account in the design of solar cells used for monitoring sun simulators, because such cells must be highly stable. In addition to the above technological measures, photon degradation can be reduced by having the base layer in the form of thin silicon wafers with a large minority-carrier diffusion length L. Large L/l_b ratios ensure effective carrier collection from the base region, even under strong illumination.

Scanning electron microscopy of solar cells

Electron microscopy and electron diffraction techniques have long attracted the attention of designers of semiconducting devices because such devices are very sensitive to phenomena occurring in the surface layer of the semiconductor. The traditional procedure in electron microscopy is to prepare replicas which are then used to investigate temperature changes in the structure of transparent conducting films of tin dioxide and antireflective zinc sulfide coatings [261,262]. On the other hand, electron diffraction has been used to examine the structure and growth of zinc sulfide films (deposited on substrates heated to different temperatures [263]) and of the photosensitive cadmium sulfide layers used in thin-film solar cells [264]. When light injection is replaced with exposure to a fine electron probe, the properties of semiconductors can be examined on a local scale (only a minor modification of the electron microscope, the specimen-holder, and the signal-amplifying system is necessary), and this substantially extends our knowledge of the structure and properties of semiconductor solar cells. For example, the ion-electron secondary emission microscope, incorporating a specially developed measuring system, has been used to investigate and visualize the potential distribution in the p—n junction and to determine the width of the space-charge region [265]. An electron beam scanned the surface of thin-film solar cells based on the copper sulfide—cadmium sulfide heterosystem (heated in air for 30 days at 120 and 200°C) and was used to determine the diffusion length of minority carriers (electrons) in copper sulfide ($L_n = 0.45 \pm 0.3\ \mu m$) and of holes in cadmium sulfide ($L_p = 0.53 \pm 0.03\ \mu m$) [266].

The advent of the scanning electron microscope and of energy analyzers capable of examining the spectrum of electrons emitted by a solid whose surface is exposed to ions, electrons, or photons, or to an electric or thermal field, has greatly extended the range of electron microscopy [267,268]. In particular, electron microscopy is now being used to investigate the surface structure and defects of single-crystal [269] and thin-film [179,188,270,271] solar cells.

The scanning electron microscope, and the various attachments used to measure the potential and current induced in the top 0.5-μm surface layer of silicon solar cells scanned with 10-50 keV electrons, have produced a large volume of information on the effect of temperature and pressure on the properties of the p—n junction and on the role of different structure defects (clusters, impurity centers, mechanical damage, and microcracks) in the deterioration in the properties of solar cells. Additional impurities in silicon (for example, nickel from the contact layer), which produce deep levels in the band gap, as well as traps and recombination centers, have also been identified. Under low levels of excitation by light on electrons, the nickel impurity produces higher sensitivity and photoconductivity, while high levels of excitation reduce the carrier diffusion length in the p-type layer [269].

It is important to note that solar cells incorporating the p—n junction that separates the charge carriers can readily be examined in the electron microscope under the conditions of induced potential and current (by analogy with U_{oc} and I_{sc} under light excitation). The effect of structure defects in uniform silicon wafers can be examined in the scanning electron microscope, but this requires the use of separating Schottky barriers produced, for example, by depositing a 400-1000 Å aluminum layer on the surface of wafers exposed to the electron beam [272].

The effect of pressure and temperature on the properties of p—n junctions and solar cells is more conveniently performed by the induced potential method because a change by 1 g in the local mechanical load on the surface of the cell, applied by a corundum needle, produces only a slight change in the induced current but an appreciable change (by 0.2 V) in the induced potential. In the induced potential method (load resistance 100 kΩ), the voltage is a linear function of temperature in the range 250-400 K, while, in the induced current method (load resistance 100 Ω), the current is practically insensitive to temperature changes in the range 250-350 K and shows a sharp reduction only at temperatures of 350-400 K [269].

The properties of thin-film solar cells based on the cadmium sulfide—copper sulfide heterosystem have been investigated in the ISM-50A scanning electron microscope, using secondary electron emission and also attachments for cathodoluminescence and X-ray microanalysis studies [250,270]. The mass number of the specimen material can be estimated quantitatively when only the high-energy component of secondary emission is recorded. Surface properties, such as the presence of dislocations, dopants, and the positions of recombination and trapping centers in the band gap of the semiconductor, can be investigated by recording the electromagnetic radiation produced by the electron beam in the specimen and by measuring the cathodoluminescence spectrum emitted by local spots with diameters of about 1 μm. Quantitative microanalysis of impurities can be performed with precision that is higher in some cases than that in others. The characteristic X-ray emission can be used in quantitative elemental microanalysis, which will reveal the presence of most of the elements in the periodic table.

The distribution of copper in different thin-film front-barrier solar cells based on the Cu_xS—CdS system produced by the "wet method," has been determined as a result of this X-ray structure analysis in the scanning electron microscope. The concentration of copper was found to be a maximum near the Cu_xS layer and then fell slowly with depth in the cadmium sulfide layer [250,270].

It is well known that copper forms deep acceptor levels in copper sulfide, the appearance of which leads to a high degree of donor compensation and an increase in resistivity, so that the long-wave impurity sensitivity of copper sulfide can increase. However, the enhanced-resistivity layer must have a relatively small depth. Studies of the current-voltage characteristics together with the cathodoluminescence and X-ray microanalysis in the scanning electron microscope have confirmed that the solar cells have the best energy parameters when copper penetrates the cadmium sulfide layer to a depth in excess of 4 μm (the total thickness of cadmium sulfide being 15-20 μm).

The orange band with a maximum at 0.598-0.605 μm in the cathodoluminescence spectra, which is due to natural defect complexes containing copper atoms and sulfur vacancies [250,270], is found to disappear when zinc sulfide and cadmium selenide are introduced into the cadmium sulfide layer. The addition of these materials results in improved-quality cadmium sulfide layers (lower departure from stoichiometry), but the attendant increase in resistivity produces an increase in the series resistance of cells and a lower efficiency. This can be overcome by additional doping of the cadmium sulfide—copper sulfide solid solutions, or by using a two-layer structure for the base region of these solar cells, namely, a high-resistivity layer consisting of pure cadmium sulfide, or a solid solution of cadmium sulfide and zinc sulfide, and a low-resistivity layer of the solid solution of cadmium sulfide and zinc sulfide (in direct contact with the molybdenum substrate). The low-resistivity layer can be produced by doping with indium [187] or zinc [273], and the high-resistivity layer by copper enrichment [179,270,273].

Two-layer structures in silicon solar cells that produce electrostatic drift fields in the doped and base regions [121,122,128-131] have already been examined in Section 2.1. In the case of thin-film solar cells, one can also have a two-layer structure not only in the base CdS and CdZnS layer but also in the thin upper layer of Cu_xS, for example, by enriching the p-type Cu_xS layers next to the heterojunction with the donor zinc impurity [187].

Auger spectroscopy is another form of electron microscopy. In 1925, Auger discovered that the excitation of the inner electron states of atoms by X-rays resulted in the emission of electrons. This phenomenon was subsequently called the *Auger effect* and the emitted electrons were referred to as Auger electrons. However, a quarter of a century elapsed before excitation by electron impact was used to ionize the inner shells of atoms, and Auger electron peaks in the secondary-electron spectra emitted by solids were recorded by differentiating the secondary-electron energy distribution [274,275]. Auger spectroscopy has turned out to be a very effective method for the elemental analysis of surfaces [275]. Successive removal of monatomic surface layers by gas, chemical, or ion etching, followed by the Auger spectroscopy of the exposed areas, can yield valuable information on the distribution of chemical elements in thin films, optical coatings, and surface layers of solar cells. Impurity concentrations of only 10^{10}-10^{12} cm^{-3} can be detected by this procedure. Its spatial resolution ranges from some tens to some hundreds of Ångstroms [267,268].

In modern Auger spectrometers, the surface of the sample is exposed to a beam of 1-5 keV electrons which ionize the inner shells of the target atoms. The resulting vacancies are filled by electrons from higher states, but the process is accompanied by the emission of a group of electrons whose energy is determined by the electronic structure of atoms in the solid. The result is that the secondary-electron energy spectrum contains Auger peaks representing the elemental composition of the solid. Auger electrons are emitted from a few monatomic layers, so that the data obtained in this way refer to the outermost surface layers.

Intermediate-energy ions (for example, Ar^+ with energies of 15 keV) are now being used to produce Auger electrons [276]. This results in higher sensitivity as compared with excitation by electron impact, and atomic and quasimolecular Auger-electron spectra can be recorded, so that we have the basic possibility of not only elemental analysis of the surface, but also of elucidating the composition of the surface layers of chemical compounds. It has been shown that, when chemical compounds are bombarded with ions, Auger processes occur in moving particles that have been displaced by the ions from their original positions by a few Ångstroms, but the Auger spectra are largely determined by the original state of the particle, so that chemical analysis of many of the compounds by Auger spectroscopy is still possible [276].

Optical scanning microscopy

Sensitive electron-microscope methods of analysis of the structure and composition of surface layers and thin films suffer from an important drawback, namely, the interaction between the electron beam and the solid may be accompanied by irreversible physicochemical processes which alter the properties of the material under investigation, and the analysis may reveal these modified properties. Optical methods of analysis, on the other hand, are less sensitive but have little effect on the nature of semiconducting materials and semiconductor devices. There is particular interest in the development of scanning optical microscopes [277,278] and their application to the study of impurities on the surface of silicon wafers [279] and semiconducting structures such as solar cells [280]. Analysis of the properties of semiconducting structures, based on the variation in the photo-emf under excitation by a narrow light probe, is a very promising development [277,278]. The photoresponse signal depends on the electrophysical and optical properties of the semiconductor, the nature of the p—n junction, and surface and other defects in the semiconductor device. By recording photoresponse signals from different points on the surface, one can determine and investigate local changes in the properties of semiconductor devices. In early work, the photoresponse signal was recorded for a fixed or a mechanically scanned light probe [31], which was a serious practical restriction.

The scanning optical microscope, on the other hand, uses electronic scanning of the light probe and the raster principle of producing the photoresponse image [277,278]. The source of light is a special electron-beam tube, and the photoresponse image is formed on the screen of a synchronous monitor. The diameter of the light probe, which governs the resolution of the scanning optical microscope, is about 2 μm.

Photographs of the photoresponse images [278] identify photolithographic defects and points of detachment of metallized contact strips from the surface of semiconductor devices. They also provide information on charge-producing phenomena in thin dielectric coatings, which give rise to the inversion of photoconductivity in the upper layers of the semiconductor. Both the scanning optical microscope and the scanning electron microscope can be used (with the necessary modifications) to record the signal from semiconducting structures under induced current or induced potential conditions.

Semiconducting structures of the p^+—n-type exposed to ammonia or iodine vapor have been investigated in the scanning optical microscope under a small negative bias [280]. Structures exposed to iodine vapor did not change their characteristics when placed in a vacuum (5.10^{-5} mm Hg), so that they can be studied by scanning electron microscopy (which differs from scanning optical microscopy by the fact that the measurements must be carried out in a vacuum). The characteristics of specimens exposed to ammonia vapor could be investigated only with the scanning optical microscope because they had to be kept in the ammonia atmosphere in air. However, when measurements were made in the scanning electron microscope, bombardment by the electron beam produced a substantial change in the reverse current in both types of specimen because of the appearance of the surface electric double layer. The scanning optical microscope was used to investigate these specimens without distorting the original structural properties, and this once again confirmed the undoubted advantages of this method in the study of complex physical and optical phenomena on solid surfaces.

3

Solar Cells with Optical Coatings

Optical coatings are used to increase the efficiency, extend the life, and improve the electrophysical and working characteristics of solar energy converters based on various physical principles, including semiconductor solar cells [23]. The problems that can be solved by applying optical coatings depend on the application of the solar cells.

When solar cells are used mainly for the conversion of solar energy directly into electric power, optical coatings reduce reflection in the region of spectral sensitivity by exploiting the interference effect (antireflective surfaces), increase the intrinsic thermal emission by the front and back sides of the cell, and protect it from damage by radiation in space and from unfavorable atmospheric conditions on the Earth.

When solar cells are placed on the exterior of thermal collectors in photothermal systems, and generate both electric and thermal power, the optical coating applied to their surfaces gives them highly unusual selective properties, namely, reduced reflection of solar radiation (and high transparency in this part of the spectrum), which leads to higher integrated solar absorption coefficient α_S, and enhanced infrared reflection, which ensures that the intrinsic thermal emission coefficient ϵ is as low as possible. Solar cells then not only generate electric power but, at the same time, act as selective optical surfaces for thermal solar-radiation collectors.

The surface of thermal collectors that is free from solar cells must carry selective black-and-white coatings with high α_S/ϵ ratio and strong absorption for solar radiation (the requirement that the coatings must be transparent to solar radiation, which must be satisfied in the case of solar cells, is then removed).

Of course, the optimum optical characteristics of coatings must be combined with radiation stability, i.e., the ability to retain the initial parameter values throughout the working life of the solar energy converter. Careful laboratory, field, and space tests on new coatings and converters carrying such coatings are essential.

3.1 Optical coatings for solar cells used as sources of electric power

The use of one- or two-layer antireflective optical coatings results in a substantial increase in the efficiency of solar cells. Theoretical estimates indicate that this increase should be 48-50%, while experiment indicates figures in the range 40-44% [7,13,23,109]. Virtually the same result is obtained by depositing more complicated (three- to seven-layer) antireflective coatings, so that these are really useful only in cascade solar cells [23], where reduced reflection must be achieved in a broad range of solar radiation wavelengths (between 0.2 and 2.5 μm) and not merely within the spectral sensitivity ranges of the individual solar cells.

For many years, most development work was directed toward improving the quality of one- and two-layer antireflective coatings deposited on the surface of solar cells in which the p—n junction was produced in homogeneous or heterogeneous materials based on silicon or gallium arsenide, or on Schottky barrier solar cells whose outer surface was coated by a semitransparent metal film. Special problems arose when reduced reflection was achieved by producing texturized surfaces on silicon cells made from single-crystal silicon [281,282] and thin-film cells based on Cu_2S—CdS and Cu_2S—$CdZnS$ heterostructures [176, 183, 186, 187, 192]. Because the parameters of antireflective coatings that are necessary for calculations are not available for new semiconducting materials, special measurements or estimates (involving comparisons of published data) have had to be made to determine the optical constants of materials such as, for example, solid solutions of aluminum in gallium arsenide, or the sulfides of copper and cadmium. Laboratory and field tests have been carried out on new types of charge-storing optical glass which, as has already been noted, can be used for protecting solar cells against damage by radiation [23,283-285].

One- and two-layer antireflective coatings and methods
of reducing the surface layer resistance of solar cells

Thin transparent films based on tin dioxide, mixtures of indium and tin oxides (ITO films), or cadmium stannates [175,218] are commonly used in solar-cell design. In early work, these films were used only as passive elements (transparent current contacts on back-barrier cells based on the copper sulfide—cadmium sulfide hetero-system [138,172,175,183]). In modern cells, for example, those based on the ITO-silicon heterosystem [217,218,223,230], the transparent oxide film acts as a wide-gap semiconducting filter-window. The separation boundary between the tin dioxide and silicon has a low rate of surface recombination, so that a transparent conducting layer can be used to produce an isotype barrier on the back surface of the two-sided solar cell [147].

Transparent conducting oxides can also be used as antireflective coatings, especially since their refractive indices lie between 1.6 and 1.9 [185,286,287] and are close to the optimum values for the materials used in one-layer coatings on solar cells without glass or polymer cover, or for the individual layers of a multilayer coating [23].

The antireflective ITO film coatings (geometric thickness 700-750 Å and surface layer resistance between 75 and 100 $\Omega/\square$) have been deposited after the very thin (approximately 900 Å) damaged surface layer of the doped region of the silicon solar cell was removed by plasma etching in silicon hexafluoride (with a radio-frequency field applied to the system). This produced a higher shunt voltage and an improved current-voltage characteristic, but was accompanied by an increase in the series resistance of the cells because of the higher surface layer resistance. Subsequent deposition of a transparent conducting ITO film coating again reduced the surface layer and series resistances, so that the efficiency of the cells increased by 43% as compared with the results obtained prior to plasma etching and the deposition of the antireflective coating [288].

The fabrication technology used for solar cells for terrestrial applications [288] includes operations such as thermodiffusion of a phosphorus impurity (to produce the p—n junction) from a film first deposited on the surface of the silicon wafers (by centrifuging), the imposition of an isotype barrier and back contact by baking on a conducting paste containing aluminum, and the application of top contacts in the form of a silver-containing paste. The silicon specimens containing the p—n junction and bearing the necessary contacts were then cut through the back side into rectangular plates by a laser beam.

Single-layer antireflective coatings produce a substantial increase in the efficiency of photocells. They are simple to make, but have one serious disadvantage. Single-layer coatings on semiconductors with a high refractive index (such as silicon) can be used to achieve virtually zero reflection at a particular wavelength, but the reflection coefficient is then found to increase rapidly with wavelength. Low reflection coefficients throughout the spectral sensitivity range of photocells can be achieved (and hence the efficiency increased to its maximum level) by using two-layer antireflective coatings [289], whose deposition can be automated just as readily as in the case of one-layer coatings [290].

By solving the equations for the reflection coefficient of a surface carrying a two-layer coating, in which the two layers have equal optical thickness, it may be shown that there are two optimum relationships between the refractive indices of the layers [291], namely,

$$n_1^2 n_3 = n_2^2 n_0 \tag{3.1}$$

and

$$n_1 n_2 = n_0 n_3, \tag{3.2}$$

where n_1, n_2, n_3, and n_0 are the refractive indices of the outer and inner antireflective films, the substrate, and the ambient medium (air), respectively.

When the optical thickness of each layer is an odd multiple of $\lambda_{min}/4$, where λ_{min} is the wavelength corresponding to the minimum reflection coefficient, the condition given by (3.1) ensures that the spectral reflection curve has a single (zero-order) minimum. When (3.2) is satisfied, the curve will have two minima, and the optical thickness of each layer is $\lambda_{max}/4$ at the maximum of the curve (which lies between the two minima).

The recurrence relations for the amplitude reflection coefficient [292-295] have been used to calculate the spectral reflection curve of silicon after the deposition of a two-layer antireflective coating consisting of zinc sulfide (n_{ZnS} = 2.3) and magnesium fluoride (n_{MgF_2} = 1.38) of different optical thickness. This was then compared with the calculated reflection coefficient of silicon after the deposition of a single-layer antireflective coating of zinc sulfide or silicon monoxide (SiO), or a two-layer coating of silicon monoxide (n_{SiO} = 1.9) and magnesium fluoride. The wavelength dependence of the refractive indices of silicon, zinc sulfide films, and silicon monoxide films was taken into account [23].

Comparison of these calculated curves shows that the widest low-reflection interval is achieved for two-layer coatings of zinc sulfide and magnesium fluoride. This is confirmed by calculations of the short-circuit current I_{sc} of solar cells under AM0 conditions, after the reflection coefficients were reduced in accordance with the calculated curves.

Multilayer coatings have shown good agreement between calculated and experimental data [296]. The most complicated problem is to produce two-layer coatings for which the external medium is not air but a protective glass or polymer layer with refractive index of 1.4-1.5. Calculations then show that it is best to take antireflective layers with refractive indices of 2.3 and 1.7 [23]. The optical thickness of the two antireflective layers must also be different: a low reflection coefficient throughout the spectral sensitivity range of the photocells is assured, for example, not only by the two-layer coating in which each layer has an optical thickness of $\lambda/4$ (where $\lambda = \lambda_{max}$ in the solar emission spectrum), but also by coatings whose thickness is not a multiple of $\lambda/4$. In particular, calculations have shown that the following two-layer coating is successful: the layer in contact with silicon has n = 2.3 and d = $\lambda/5$, while the second layer has n = 1.7 and d = $\lambda/3$. The two-layer coating is protected by glass or transparent polymer film with n = 1.4-1.7 and a "noninterference" thickness (between 50 and 150 μm).

By applying an ohmic contact to the highly doped upper layer of the semiconductor photocell through the antireflective coating [23,78,79,297], we can minimize the danger of short-circuiting photocells with a shallow p—n junction, and sharply reduce leakage currents. This method of contact-layer deposition is complicated in the case of the two-layer antireflective coating because of its substantial total thickness as compared with the single-layer coating. This deficiency of two-layer antireflective coatings can be removed by using a transparent conducting material such as, for example, an ITO film for one or both layers.

Two-layer coatings with low reflection throughout almost the entire spectral sensitivity range of photocells have been produced [289]. The first layer of the coating is a zinc sulfide film deposited by thermal evaporation in vacuum, while the second layer is an ITO film produced by evaporation by electron bombardment in a vacuum, followed by annealing in air at 200-250°C, or by ion plasma deposition from a compacted ITO tablet. The surface layer resistance of ITO films with the geometric thickness l necessary for two-layer antireflective coatings (1100-1200 Å) is found to be between 80 and 100 $\Omega/\square$. Figure 3.1 (curves 3 and 5) shows the reflection coefficient of the surface of silicon solar cells with a two-layer antireflective coating (ZnS-ITO) before and after the application of a protective glass plate. For comparison, we also show the reflection curves for silicon without the antireflective coating (curve 1) and for silicon with a single-layer antireflective coating (curves 2 and 4).

Two-layer coatings of this type not only ensure a low reflection coefficient of the silicon surface in a wide spectral range, but also reduce the surface layer resistance by a factor of more than two, which means that the separation between contact strips on the front surface can be increased and the efficiency can be raised by reducing the area occupied by the front contact.

Optical characteristics similar to those shown in Fig. 3.1 are obtained when the zinc sulfide layer in the single-layer and two-layer antireflective coatings is replaced with cerium, titanium, or tantalum films.

Solar cells with textured surface

The textured surface produced after the application of selective etching solutions to silicon wafers oriented in the {100} plane ensures that the subsequent deposition of even single-layer antireflective coatings results in a very low (1-2%) reflection coefficient

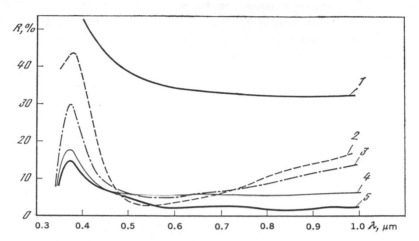

FIG. 3.1. Reflection coefficient of the surface of a silicon solar cell as a function of wavelength for different antireflective coatings: 1) no coating; 2) single-layer ZnS film (l = 630 Å); 3) two-layer ZnS coating (l = 520 Å) + ITO (l = 1170 Å); 4, 5) protective glass on a surface coated with a single-layer antireflective coating (4) and a two-layer antireflective coating (5).

throughout the spectral sensitivity range of silicon solar cells [281]. This method of reducing the reflection coefficient has the following disadvantage: the surface absorbs beyond the fundamental absorption edge, so that the nonphotoactive part of the solar spectrum between 1.1 and 2.5 μm cannot be transmitted by the cell [23].

However, the textured surface ensures that, owing to multiple reflections, all the photoactive solar radiation is absorbed by the semiconductor even in the case of very thin wafers. The thin wafers necessary for producing silicon solar cells with high power-to-mass ratio can be obtained by etching single-crystal silicon wafers parallel to the {100} crystallographic plane in a boiling 20% solution of KOH or NaOH in water [298]. The rate of etching is then 5-7 μm/min, so that ordinary wafers with an initial thickness of 200-250 μm, placed vertically in fluoroplastic containers, are transformed into thin, almost film-like layers of 40-50 μm in about 20-30 min. These layers are flexible but strong. All subsequent operations are performed as usual.

The thermodiffusion of dopants can be performed by any of

the well-known methods, such as diffusion in a gas flow, diffusion from thin films or through a porous oxide film, or diffusion from a solid source such as a ceramic or a glass containing phosphorus or boron. The isotype barrier is produced not only by the thermo-diffusion of boron (for base layers of n-type silicon), but also by depositing aluminum in the form of a paste containing an organic binder, or by thermal evaporation in a high vacuum, followed by the baking-on of the deposited layer at 800°C for 10 min in a furnace flushed with argon.

It is found that the surface topology of thin wafers produced in this way depends on the temperature of the solution: when the temperature is raised above 100°C, the surface is smooth and gleaming but, when the temperature is less than 100°C (or the concentration of KOH is less than 15-18%), the surface is textured and covered by pyramidal pits on the {100} plane.

The power-to-weight ratio is up to 1-1.5 kW/kg [140-142] in the case of high-efficiency solar cells made from thin silicon wafers, where the high efficiency is achieved by using n^+-p-p^+ structures and multiple transmission of light reflected by coatings on the back side of the wafers, which is free of contacts.

Antireflective metal film on the surface of a solar cell

Schottky-barrier solar cells, semitransparent metal layers in the selective insulation of thermal collectors, and the surfaces of cells used in composite photothermal converters must all be coated with antireflective nonabsorbing dielectric layers. However, the choice of the materials for these layers is so restricted that the problem can be solved by choosing the metal layer material for the given antireflective coating and not the other way around. Of course, the first step is to determine the combination of materials and the optical constants and layer thicknesses for which the minimum reflection coefficient will be attained [23,299]. From the technological point of view, it is obviously more convenient to use a single-layer antireflective coating.

The electromagnetic theory of light gives the most complete and accurate description of optical effects in thin-film multilayer systems [291,295]. The reflection, transmission, and absorption coefficients of optical systems can be calculated from this theory in a relatively simple way. The calculation reduces to the solution of a boundary-value problem, i.e., to the determination of the amplitudes of the electric and magnetic fields on all boundaries of the multilayer system illuminated by a given light wave. The recurrence relations for the complex amplitude reflection and transmission coefficients of the $(j - 1)$-th separation boundary at normal incidence are as follows:

$$r_{j-1} = \frac{f_{j-1} + r_j \exp(-2i\Phi_j)}{1 + f_{j-1} r_j \exp(-2i\Phi_j)}, \qquad (3.3)$$

$$t_{j-1} = \frac{g_{j-1} t_j \exp(-i\Phi_j)}{1 + f_{j-1} r_j \exp(-2i\Phi_j)}, \qquad (3.4)$$

where

$$f_{j-1} = (N_{j-1} - N_j)/N_{j-1} + N_j) \qquad (3.5)$$

and

$$g_{j-1} = 2N_{j-1}/(N_{j-1} + N_j) \qquad (3.6)$$

are the classical Fresnel coefficients for the $(j - 1)$-th boundary, $N_j = n_j - ik_j$ is the complex refractive index of the j-th layer, $\Phi_j = 2\pi N_j l_j/\lambda$ is the phase thickness of the j-th layer, and l_j is the geometric thickness of the j-th layer. The conditions $r_m = f_m$ and $t_m = g_m$ are satisfied on the separation boundary between the last (m-th) layer and the semi-infinite substrate. The recurrence procedure begins with the last layer in the system and ends with the determination of r_0 and t_0. The reflection and transmission coefficients of the system are then given by

$$R = |r_0|^2,$$
$$T = |t_0|^2 \mathrm{Re}(N_{m+1})/n_0, \qquad (3.7)$$

where n_0 is the refractive index of the medium from which the radiation arrives (nonabsorbing medium) and $\mathrm{Re}[N_{m+1}]$ is the real part of the complex refractive index of the semi-infinite substrate.

For an optical structure consisting of a transparent single-layer antireflective film of a dielectric on a metal substrate with a complex refractive index, (3.3)-(3.7) yield

$$r_0 = \frac{f_0 + r_1 \exp(-i4\pi n_1 l_1/\lambda)}{1 + f_0 r_1 \exp(-i4\pi n_1 l_1/\lambda)}$$

$$f_0 = \frac{n_0 - n_1}{n_0 + n_1}, \qquad r_1 = f_1 = \frac{n_1 - n_2 + ik_2}{n_1 + n_2 - ik_2},$$

$$R = \frac{|f_0|^2 + |f_1|^2 + 2|f_0| |f_1| \cos(4\pi n_1 l_1/\lambda + \arg f_0 - \arg f_1)}{1 + |f_0|^2 |f_1|^2 + 2|f_0| |f_1| \cos(4\pi n_1 l_1/\lambda - \arg f_0 - \arg f_1)}$$

where

$$|f_0|^2 = \frac{(n_0 - n_1)^2}{(n_0 + n_1)^2}, \qquad |f_1|^2 = \frac{(n_1 - n_2)^2 + k_2^2}{(n_1 + n_2)^2 + k_2^2}, \qquad (3.8)$$

$\arg f_0 = \pi$ since $n_1 > n_0$ (for air $n_0 = 1$), and

$$\arg f_1 = \text{arc tan} \frac{2n_1 k_2}{n_1^2 - n_2^2 - k_2^2}. \qquad (3.9)$$

The condition for the thickness l_1 of the film which, for given values of the optical constants n_1, n_2, and k_2 and wavelength λ, will ensure the minimum reflection coefficient of the system

$$R_{\min} = (|f_0| - |f_1|)^2 / (1 - |f_0| |f_1|)^2, \qquad (3.10)$$

is

$$4\pi n_1 l_1/\lambda = \arg f_1 + 2\pi m \qquad (m = 0, 1, 2, \ldots).$$

When m = 0, we have

$$n_1 l_1 = {}^1\!/_4 \lambda \arg f_1/\pi, \qquad (3.11)$$

and, according to (3.9),

$$n_1 l_1 = {}^1\!/_4 \lambda \text{ arc tan} \left(2n_1 k_2/(n_1^2 - n_2^2 - k_2^2)\right)/\pi, \qquad (3.12)$$

from which it is clear that, when $k_2 = 0$ and $n_1 < n_2$, the optical thickness of the film must be equal to $\lambda/4$ (this result is well known for an antireflective coating on nonabsorbing surfaces), but an antireflective dielectric film cannot be produced for $n_1 > n_2$.

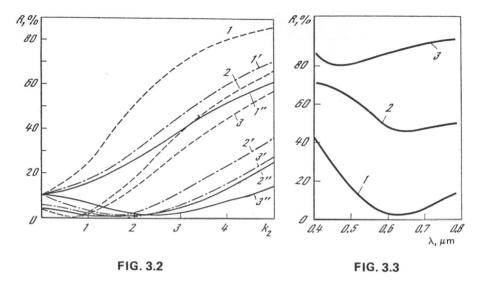

FIG. 3.2   FIG. 3.3

FIG. 3.2. Reflection coefficient of a structure consisting of an antireflective coating on metal as a function of the absorption index of the metal for different values of the refractive index of the film (n_1) and metal (n_2): 1, 1', 1") n_2 = 0.5; 2, 2', 2") n_2 = = 1.5; 3, 3', 3") n_2 = 2.0; 1-3) n_1 = 1.45; 1'-3') n_1 = 2.3; 1"-3") n_1 = 2.8.

FIG. 3.3. Reflection coefficient of opaque films of nickel (1), aluminum (2), and silver (3), covered with an antireflective coating of zinc sulfide of different thickness: 1, 3) d = 350 Å; 2) d = = 485 Å.

A number of absorbing materials with different refractive indices can be chosen for the antireflective film. For example, one can take refractive indices of 1.45 (SiO_2), 2.3 (ZnS), and 2.8 (TiO_2, SiC).

Equation (3.12) is used for each fixed value of n_1 to construct a family of curves representing the optimum optical thickness d = = $n_1 l_1$ of the antireflective film as a function of the substrate absorption coefficient k_2 for a number of values of the substrate refractive index n_2 [23,299]. The data obtained in this way show that, in the case of metals, the optimum thickness of the coatings must be much less than $\lambda/4$.

We must now determine the ratios of optical constants for which zero reflection is produced. When (3.11) is satisfied, we have

TABLE 3.1
Effect of an Antireflective Transparent Zinc Sulfide
Film on a Metal Layer

Metal	λ_{min}, μm	R, %	l_{ZnS}, nm	R_{min}, %	$n_2\lambda_{min}$	$k_2\lambda_{min}$	d_{ZnS} Exp.	d_{ZnS} Calc.
Ag	0.480	95.0	38.0	79.5	0.30	4.71	$\lambda/5.49$	$\lambda/5.62$
Al	0.650	84.5	48.5	46.0	1.08	4.85	$\lambda/5.83$	$\lambda/5.50$
Ni	0.625	38.5	35.0	1.0	1.54	1.89	$\lambda/7.76$	$\lambda/7.62$

$R_{min} = 0$ for $|f_0| = |f_1|$ [since R_{min} is given by (3.10)] and, if we take into account (3.8), we find that this occurs for

$$k_2 = ((n_2-n_0)(n_1{}^2-n_0 n_2)/n_0)^{1/2}. \tag{3.13}$$

When $k_2 = 0$, we have the usual antireflection conditions $n_1{}^2 = n_0 n_2$ for $d = \lambda/4$.

When $k_2 \neq 0$, zero reflection will occur if the following conditions are simultaneously satisfied:

1) $n_2 > n_0$ and $n_1 < \sqrt{n_0 n_2}$, where the optical thickness of the antireflective layer satisfies the condition $n_1 l_1 = \lambda \arg f_1 /4\pi$

2) $n_2 < n_0$ and $n_1 < \sqrt{n_0 n_2}$, which is impossible because $n_0 = 1$ and n_1 can only be greater than unity.

Figure 3.2 shows the results of calculations based on (3.10) and (3.11). Zero reflection corresponds to values of k_2 for which (3.13) is satisfied. The condition for zero reflection is

$$n_1 = (n_0(n_2+k_2{}^2/(n_2-n_0)))^{1/2}.$$

This is valid for $n_2 > n_0$ and arbitrary k_2. When $n_2 < n_0$ and $k_2 \neq 0$, the value $R_{min} = 0$ is impossible.

The results shown in Fig. 3.2 enable us to derive certain preliminary conclusions about the effectiveness of antireflective coatings with different refractive indices. The most successful is the coating with maximum possible refractive index because, for fixed n_2 and k_2, this produces the lowest reflection coefficient. Moreover, such coatings have a wide reflection coefficient minimum, i.e., they are less sensitive to deviations of n_2 and k_2 from optimum values.

These conclusions have frequently been confirmed experimentally. For example, the deposition of an antireflective coating of zinc selenide with n = 2.3 on semitransparent nickel, aluminum, or silver layers results in a much higher transparency [23] than the deposition of a silicon monoxide film with n = 1.8-1.9 [219]. As already noted, metals differ from dielectrics in that, in their case, the optimum optical thickness of the coating is less than $\lambda/4$.

The above results have been confirmed experimentally. Semitransparent metal films of silver, aluminum or nickel were deposited on glass plates by evaporation in a vacuum, and a film of zinc sulfide was deposited on top of them. Its thickness was monitored with a quartz resonator, forming part of the measuring system. Figure 3.3 shows the reflection coefficient of the resulting structures, and Table 3.1 lists values of the optical thickness of the zinc sulfide film (n_{ZnS} = 2.3) together with the calculated values based on (3.11).

Since the optical constants of metal films, especially silver films, may depend on the deposition conditions, the first step was to determine the constants n_2 and k_2 of these films. This required the simultaneous solution of (3.10) and the equation

$$R = \frac{(n_2 - 1)^2 + k_2^2}{(n_2 + 1)^2 + k_2^2}.$$

The former gives the minimum reflection coefficient of a metal surface coated with an antireflective film, and the latter gives the reflection coefficient of the same metal without the coating at the wavelength λ_{min} corresponding to R_{min} (R_{min} and R were determined experimentally).

Of all the metals examined, only nickel was found to have suitable optical constants that could be exploited to produce zero minimum reflection coefficient by depositing an antireflective coating [see Eq. (3.13)].

The maximum antireflective effect can thus be attained by coating the collector or solar cell with a top layer consisting of a metal such as nickel, iron, or titanium, whose optical constants are close to those necessary to produce the theoretical zero reflection coefficient. This conclusion is also valid for multilayer interference coatings containing semitransparent metal films.

It is important to note that the optical properties of semi-transparent nickel and titanium films approach those of semiconducting layers [286,294], and can be used as antireflective coatings on silicon solar cells with a Schottky barrier consisting of a highly reflecting metal such as aluminum [220]. The optimum thickness of the aluminum barrier is chosen on the basis of the measured spectral reflection and transmission coefficients, and the layer resistance of the metal film on glass and silicon, which are found to be 200 $\Omega/\square$ for a 50-Å aluminum layer.

A 30-Å titanium film deposited on top of the aluminum layer acts as an antireflective coating and produces an increase in the transparency of the aluminum layer by between 29 and 61% at $\lambda = 0.9$ μm.

Antireflective coatings on heterojunction solar cells

The uppermost layers on heterojunction solar cells — the ITO—Si, Cu_2S—CdS, AlGaAs—GaAs film — perform two functions, namely, they act as the photoactive region in which excess carriers are produced by the short-wave part of the solar spectrum, and as the current-conducting electrodes. Since the refractive index of ITO, Cu_2S, and AlGaAs lies between the refractive index of air (or glass or polymer protecting layer) and the base layer of the solar cell (usually silicon, cadmium sulfide, or gallium arsenide cell), these top layers on heterocells can also act as efficient antireflective coatings. The wavelength dependence of the reflection coefficient of the surface of heterojunction solar cells is found to show a well-defined minimum due to the antireflection effect [23].

Calculations and experiment have shown that the deposition of an antireflective film with n = 1.7-1.8 (for example, Al_2O_3) on top of the ITO, Cu_2S, AlGaAs layers, and the optimization of the thickness of the uppermost layer of the heterojunction, result in a two-layer antireflective coating which reduces reflection from the cell covered with glass or a polymer layer with n = 1.5 virtually throughout the spectral sensitivity range of heterojunction solar cells. In the absence of the uppermost protective glass or polymer (boundary with air for which n = 1), the optimum top antireflective layer can be a film of silicon dioxide (n = 1.44-1.45) or magnesium fluoride (n = 1.38). Calculations are found to be in reasonable

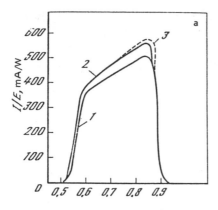 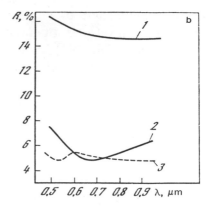

FIG. 3.4. Spectral sensitivity (a) and reflection coefficient (b) of AlGaAs–GaAs heterojunction solar cells for different antireflective coatings: 1) without the antireflective coating; 2) single-layer anti-reflective coating (optimum parameters n_1 = 2.1, l_1 = 800 Å); 3) two-layer coating (n_1 = 1.7, l_1 = 1000 Å — outer layer; n_2 = 2.3, l_2 = 750 Å — inner layer).

agreement with experiment [23]. A still broader low reflection interval is produced by depositing a two-layer antireflective coating on the uppermost layer of the heterojunction cell.

One- and two-layer antireflective coatings, producing reduced reflection at the boundary between the surface of the Cu_2S–CdS or AlGaAs–GaAs heterojunction cell and an ambient medium with n = 1.5, have been optimized by calculation [300]. The Cu_2S and AlGaAs layers were regarded as "noninterference" layers in these calculations. Figure 3.4 shows the numerical optimization performed for the AlGaAs–GaAs heterosystem. The measured initial spectral sensitivity of the AlGaAs–GaAs solar cells was used in these calculations (Fig. 3.4a, curve 1).

The spectral sensitivity was recalculated with allowance for the change in the wavelength dependence of the reflection coefficient after the deposition of the antireflective coatings, using a special program based on the well-known recurrence relations given in [191-195]. This involved averaging over the phase thickness of the "noninterference" AlGaAs layer, using formulas analogous to those derived for multilayer structures with "noninterference" layers of glass, transparent rubber, and silicon [23]. There is no doubt that,

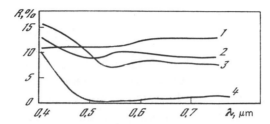

FIG. 3.5. Reflection coefficient of thin-film Cu_2S- $Zn_xCd_{1-x}S$ heterojunction solar cells: 1, 2) Cu_2S layer produced by reaction in the solid phase and chemical treatment, respectively; 3, 4) after the deposition of antireflective ZnS coating (d = 0.12-0.13 μm) and SiO_2 (d = 0.13-0.16 μm) on the Cu_2S layer produced by chemical treatment, respectively.

in the case of AlGaAs layers of optimum interference thickness, the deposition of one- or two-layer antireflective coatings should lead to a substantial reduction in the reflection coefficient and an increase in spectral sensitivity.

It is important to note that the use of the upper, photoactive region of the heterojunction as one of the layers of an antireflective multilayer coating becomes appreciably less effective when the refractive indices of the upper and base layers of the heterojunction solar cells are either close to one another or even equal. Thus, when zinc sulfide is added to the cadmium sulfide base layer of thin-film heterojunction solar cells with a copper sulfide top layer, it is found that the band gap of the base layer increases and its refractive index decreases to a value characteristic for copper sulfide. This is responsible for the absence of the interference minimum on the wavelength dependence of the reflection coefficient of thin-film solar cells containing the $Cu_2S-Zn_xCd_{1-x}S$ structure [187]. When an antireflective zinc sulfide film with a refractive index close to that of copper sulfide and of the $Zn_xCd_{1-x}S$ base layer is deposited on the outer surface of a cell by evaporation in a vacuum, this does not result in the appearance of a well-defined interference minimum on the wavelength dependence of the reflection coefficient (Fig. 3.5, curve 1), or in an improvement in the properties of the cell. However, the deposition of the silicon dioxide antireflective coating does result, even in this case, in a reduction in the reflection coefficient of the surface of a heterojunction solar cell (curve 4).

Charge-retaining glass as a protective cover
against radiation damage

A considerable protecting effect can be achieved with transparent covers of fursed quartz or nondarkening glass on solar cells and batteries working in the Earth's radiation belts [22,23,96]. By absorbing or attenuating the low-energy proton and electron components of radiation belts, relatively thin (0.3-0.5 mm) glass covers produce an appreciable reduction in the number of particles entering the solar cell, and thereby reduce radiation damage by a substantial factor. This extends the life of cells and batteries in the presence of high-intensity corpuscular radiation.

Further protection against radiation in space can be achieved by using glass capable of accumulating electric charge when exposed to radiation-belt electrons. Terrestrial studies have shown that the radiation dose behind a charged glass cover is reduced by 10-40%, depending on the ratio of the range of the electrons to the thickness of the cover. The electron dose is reduced because the electrons interact not only with the material of the cover but also with its internal electric field [283].

Charge collection by coatings under the conditions prevailing in the Earth's radiation belts can be exploited to increase the efficiency of radiation shielding without the necessity for preliminary charging of the glass on Earth.

Inorganic phosphate glass [284] is one of the dielectrics with the necessary thermal, radiation, and optical stability, which are capable of producing strong internal electric fields under bombardment by electrons with a long (tens or hundreds of hours) relaxation time. Maximum back scattering of electrons is produced by phosphate glass with effective atomic number $Z_{eff} = 8$ and density 2.5 g//cm^3 [285].

Optical covers for "electromagnetically clean"
solar batteries

The electromagnetic field produced by the solar battery can often itself distort the performance of the many probes and instruments carried by spacecraft, including those designed to measure the charged-particle distribution in the Earth's radiation belts. This is a serious scientific problem [21]. A number of transparent electrically conducting coatings have now been developed and can be deposited in various ways on thin glass plates, including glass covers for solar batteries. They consist of wide-gap doped semiconducting oxides, such as tin dioxide, indium trioxide, and their mixtures [218], and are deposited in the form of three-layer structures (dielectric + semitransparent metal + dielectric [23]). On the outer surface of protective glasses, it is better to deposit the light, stable, and mechanically strong coatings of wide-gap semiconductor oxides. The most complicated task that arises in relation to this unusual scientific and technological problem is the optimization of the thickness of these coatings. The surface layer resistance is a further problem. Well-conducting thick films of such oxides produce high reflection in the infrared and low integrated thermal emission coefficient ϵ, which sharply increases the working temperature of solar batteries, and leads to a reduction in efficiency. At the same time, the increase in the surface layer resistance produces a deterioration in the shielding properties of covers and complicates the development of an electromagnetic shielding system.

Measurements of the integrated optical coefficients of transparent conducting coatings deposited in various ways on the outer surface of the protective glass of solar batteries were performed at room temperature, using the FM-59 photometer and the FM-63 thermoradiometer, while the surface resistivity ρ_l was measured by the probe method (Table 3.2). When the surface resistivity of the conducting coating was between 1.0 and 2.0 kΩ/$\square$, the emissivity ϵ rose to a value typical for quartz glass [21].

Measurements of the specular reflection coefficient as a function of wavelength (Fig. 3.6) clearly show that the low emissivity ϵ corresponding to high electrical conductivity of oxide coatings is due to the metallic character of the reflection. Experiment shows that, when the surface resistivity of the coatings is between 1.0 and 2.0 kΩ/$\square$, it is possible to produce a reliable electrostatic shielding system which will remove the electric component of the electromagnetic field generated by the solar batteries themselves [301]. The integrated thermal emission coefficient of the surface of solar cells is then 0.8-0.82, which means that the working temperature of "electromagnetically clean" batteries can be held at a sufficiently low level.

TABLE 3.2
Parameters of Transparent Electrically Conducting Coatings
on Radiation-Shielding Glasses for Batteries

Method of deposition	ρ, kΩ/$\square$	ε	α_S
$10\% SnO_2 + 90\% In_2O_3$			
Ion-plasma deposition in argon	0.02	0.24	0.84
	0.1	0.52	0.87
	1.2	0.82	0.91
Electron-beam evaporation in a vacuum followed by annealing	0.05	0.31	0.85
SnO_2			
Chemical pulverization	0.6	0.64	0.89
	2.0	0.82	0.91
Plasma-ion sputtering in argon	0.7	0.78	0.9
In_2O_3			
Plasma-ion sputtering in argon	1.0	0.80	0.91

Note. The coating parameters were measured for shielding glasses attached to the surface of the solar cell with transparent silicone rubber [5].

3.2   Solar cells used in composite photothermal systems

One of the most promising solar-energy converters is the composite photothermal system which generates both thermal and electric power, and has a higher efficiency than that of the thermal collector. Its efficiency is also higher than the sum of the efficiencies of the thermal and photoelectric systems working separately. Efficient photothermal converters can be based on solar collectors in which semiconductor photocells are placed on the absorbing surface instead of solar coatings [302,303]. Photothermal systems can employ flat collectors, but higher solar energy conversion efficiencies and cheaper systems can be attained by using the more effective evacuated solar collectors working together with solar concentrators [23].

Because of the high power output of composite photothermal converters, they are very effective in domestic heating. They also have certain specific applications, for example, in the temperature-regulating systems of spacecraft.

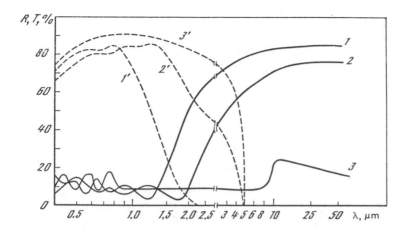

FIG. 3.6. Reflection (1-3) and transmission coefficients (1'-3') of shielding glass for solar cells coated with transparent ITO films (10% SnO_2 —90% In_2O_3) deposited by the plasma-ion method. Surface resistivity: 1, 1') 0.02 kΩ/□; 2, 2') 0.1; 3, 3') 1.0 kΩ/□.

Silicon photocells, whose fabrication technology is now well established, can be used in photothermal systems. One of the factors restricting the efficiency of planar photocells working with concentrated radiation is the power loss in the spreading resistance of the doped layer and in the resistance of contact strips on the working surface. However, the design of photocells for photothermal systems demands the optimization of not only the electrophysical but also optical parameters. Even without the application of special coatings, the photocell surface has selective properties (the antireflective polished surface of a highly doped silicon wafer has an integrated solar absorption coefficient $\alpha_S \cong$ 0.9-0.92 and integrated thermal emission coefficient $\epsilon \cong$ 0.19-0.24), but only for relatively thick doped layers ($l_d \cong$ 2-3 μm) [23,109].

For solar cells used in photothermal systems, the contact grid configuration on the working surface must be such as to reduce to a minimum the spreading resistance of the thin (0.3-0.5 μm) doped layer, and the optical coatings must be capable of producing $\epsilon \cong$ 0.1 [304]. The surface is then an efficient absorber of solar radiation, not only in the spectral sensitivity range of the photocell, but also beyond the fundamental absorption edge of silicon [305].

Optical properties of coatings for solar cells
in photothermal systems

Two groups of coating based on semiconducting metal films and dielectric layers of "interference" thickness have been developed for the solar cells used in photothermal systems or collectors [306].

Selective coatings in the first group are designed for solar cells working in concentrated sunlight [304]. They are deposited on the front surface of the usual solar cells with a solid back contact. In addition, the coatings ensure the necessary optical properties in a broad spectral range, and reduce the series resistance of solar cells, which is a necessary prerequisite for the efficient conversion of solar energy into electric power in concentrated sunlight.

Selective coatings in the second group were developed for solar cells transparent to the infrared [23,110,111]. They are deposited on the back surface of these cells and increase the absorption coefficient α_S (as well as the thermal utilization of solar radiation beyond the fundamental absorption edge of the semiconductor). They also reduce the intrinsic thermal emission coefficient of the cells. Solar cells with this type of coating have good electrical characteristics under illumination by direct sunlight.

Experimental work on the development of selective multilayer coatings has been based on the results of numerical optimization (see Section 3.1 and [23,111,304]).

Selective coatings in the first group were deposited on solar cells with p–n junction depth $l_d \cong 0.8$ μm. The doped layer was produced by thermodiffusion through a porous oxide film, and a stepped impurity distribution was established [121,122,125]. The coatings were deposited by thermal evaporation in a vacuum at a pressure of 10^{-3} Pa. The cells were protected from shorting during the metal deposition process by a mask over their edges. The development of such coatings necessitated a large number of experiments because it was essential to achieve not only the necessary optical parameters but also to ensure good ohmic contact between the coating and the surface of the solar cell, as well as low surface resistivity of the thin metal layer that is part of the solar cell.

The structures that were investigated are shown schematically in Fig. 3.7. The initially transparent silver film with a thickness of

$l \cong 70$ Å was deposited directly on the surface of the solar cell and was followed by the antireflective zinc sulfide layer of thickness $l \cong 500$ Å (Fig. 3.7a). The glass plate was inserted into the vacuum chamber, and the thickness of the metal film was monitored during deposition by measuring the transmission of the silver plus glass system at $\lambda = 0.65$ μm. A silver film with $l \cong 70$ Å on a glass substrate had a transmission coefficient $T \cong 60\%$. This figure is close to the data published in [293,294]. The surface resistivity of the film, measured by the four-probe method, was approximately 5 $\Omega/\square$. The series resistance of solar cells coated in this way, determined by the method proposed in [71], was lower but was not less than 0.5 $\Omega \cdot cm^2$.

A series resistance of less than 0.5 $\Omega \cdot cm^2$ was attained for silicon solar cells by depositing a thicker silver film ($l \cong 90$ Å). The corresponding transition coefficient of the silver–glass system was $T \cong 40\%$ [293,294] and the surface resistivity was $\rho_l \cong 2$ $\Omega/\square$. The same goal was also achieved by first depositing on the surface of the solar cell a thin chromium or titanium sublayer (transmission coefficient on glass $T \cong 90\%$; see Fig. 3.7b and c). The system was finally annealed in vacuum at 200-220°C for 0.5 hr. This procedure was used in subsequent depositions of coatings of this type.

The solar cell illustrated in Fig. 3.7a and d has a ZnS film ($l \cong 500$ Å) deposited directly on the silicon surface. Apart from solar cells, doped silicon disks without contacts were used as substrates for the deposition of these layers, and were subsequently used as controls in reflection coefficient measurements.

Figure 3.8 shows the reflection coefficient of selective optical coatings based on zinc sulfide + silver + zinc sulfide + silicon and zinc sulfide + silver + silicon structures with the same thickness of silver (the transmission coefficient on glass was approximately 40%). The shape of these curves in the infrared is an indication of the electrical properties of the silver film. The high reflection coefficient of the surface coated with zinc sulfide + silver + zinc sulfide combination shows that the silver film deposited on the zinc sulfide layer has better electrical conductivity than the silver film deposited directly on the silicon surface. Measurements actually show that $\epsilon = 0.1$ for the zinc sulfide + silver + zinc sulfide + silicon structure, whereas $\epsilon = 0.6$ for the zinc sulfide + silver + silicon structure. The difference in the electrical conductivity of the films is explained by the more regu-

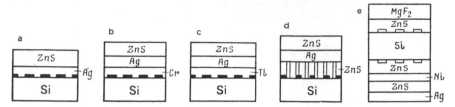

FIG. 3.7. Configuration of selective solar cells for photothermal systems and collectors: a-d) conventional cells with solid back contact; e) configuration transparent in the infrared.

lar structure of the silver film in the zinc sulfide + silver + zinc sulfide + silicon system. A similar phenomenon was noted during the deposition of single-layer films of silver and silver + zinc sulfide structures on glass [307,308]. The silver film deposited by evaporation in a vacuum onto the silicon or glass surface without the zinc sulfide sublayer is found to contain cracks and breaks.

When the zinc sulfide + silver + zinc sulfide + silicon structure is produced, the regular porosity of the zinc sulfide layer deposited directly on the silicon surface produces the contact between the silver film and silicon, and this is confirmed by the low series resistance of solar cells of this structure ($0.2\text{-}0.5\ \Omega\cdot\text{cm}^2$).

Thus, selective zinc sulfide + silver + zinc sulfide coatings deposited on the front surface of silicon solar cells ensure that $\epsilon \cong 0.1$ and that the short-circuit current is greater by 15% (because of the partially antireflective silicon surface in the spectral sensitivity range of the cell). The silver film produces a relatively rapid rise in the reflection coefficient of the structure in the near-infrared and, hence, an incomplete absorption of solar radiation (most of which is concentrated in the region $\lambda < 4\ \mu\text{m}$), while some absorption of light, which heats the silver film, is not in itself an undesirable effect in the composite photothermal solar energy converter.

The p-type silicon with resistivity $\rho_{Si} = 1$ and 25 $\Omega\cdot\text{cm}$ was used in prototype multilayer selective coatings in the second group, in which the layer disposition was as shown in Fig. 3.7e. The same silicon was also used to produce p—i—n and n^+—n—p^+ solar cells. The thickness of the polished silicon wafers was 0.3 mm, and the thickness of the other layers in the structure was close to the calculated optimum value [23]. The coatings were produced by vacuum evapo-

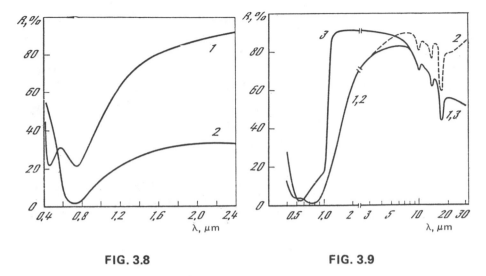

FIG. 3.8   **FIG. 3.9**

FIG. 3.8. Reflection coefficient of different selective optical coatings on doped silicon wafers with a p–n junction: 1) ZnS–Ag–ZnS; 2) ZnS–Ag.

FIG. 3.9. Reflection coefficient of silicon wafers with a reflecting back coating of opaque silver and the following selective coatings: 1, 2) MgF$_2$–ZnS–Si–ZnS–Ni–ZnS [1] 1) ρ_{Si} = 1 Ω·cm; 2) ρ_{Si} = 25 Ω·cm] ; 3) ZnS–Si (ρ_{Si} = 1 Ω·cm).

ration, and the layer thickness was monitored by a calibrated quartz resonator.

Figure 3.9 shows the reflection coefficient of structures obtained in this way, in which the nickel layer thickness was about 100 Å and the highly reflecting metal was silver (curves 1 and 2). For comparison, we also show the results for the structure with a single-layer antireflective coating of zinc sulfide and an opaque silver layer deposited directly on the back surface (curve 3).

The proposed coatings (curves 1 and 2) ensure that the integrated absorption coefficient is $\alpha_S \cong 0.90$ under AM2 and AM1.5 conditions. However, when silicon with ρ_{Si} = 1 Ω·cm (curve 1) is employed, the reflection coefficient falls sharply for $\lambda > 8$ μm, so that ϵ is 0.34 at room temperature. Infared measurements of the coefficients of transmission and reflection of polished silicon wafers, 0.3 mm thick (without coatings), have shown that this reduction

in R is due to an increase in the absorption coefficient of silicon
(A = 1 − R − T).

Calculations performed by I. P. Gavrilova can be used to estimate the reduction in the reflection coefficient as the thickness of
the silicon wafer (ρ_{Si} = 1 Ω·cm) is reduced from, say, 0.3 to 0.1
mm. To determine the infrared absorption index k of silicon, we
used the well-known refractive index of this material n = 3.42 [25,
27,28] and one of the following formulas for the air-silicon-air
optical system:

$$R = 1 - \frac{(1-r)(1 - r\exp(-8\pi kl/\lambda))}{1 - r^2 \exp(-8\pi kl/\lambda)} \qquad (3.14)$$

or

$$T = \frac{(1-r)^2 \exp(-4\pi kl/\lambda)}{1 - r^2 \exp(-8\pi kl/\lambda)} . \qquad (3.15)$$

These were obtained by adding the intensities of multiply reflected
rays (the silicon wafer has a "noninterference" thickness l), remem-
bering that the intensity is reduced by the factor exp(−4πkl/λ) on
each traverse of the wafer. The quantity r in (3.14) and (3.15) is
given by

$$r = |f_0|^2 = |f_1|^2 = \frac{(n-1)^2 + k^2}{(n+1)^2 + k^2} ,$$

where $|f_0|^2$ and $|f_1|^2$ are the Fresnel reflection coefficients of the
air—silicon and silicon—air boundaries. The values of R and T were
deduced from measurements on silicon wafers with l = 0.3 mm.

The calculated values of k at λ = 10, 20, 25, and 30 μm are
0.0004, 0.0013, 0.0021, and 0.027, respectively. The absorption
coefficient A = 1 − R − T was calculated from (3.14) and (3.15)
for l = 0.1 mm at λ = 10, 20, 25, and 30 μm. It was found to fall
from 11.5, 23.5, 25.0, and 26.5% (the experimental values for l =
= 0.3 mm) to 4.9, 7.6, 9.6, and 10%, respectively. It is thus clear
that a reduction in the thickness of the silicon wafer on which the
silver coating is deposited produces a substantial reduction in the
value of ε.

High-resistivity silicon has lower absorption in the infrared
[25,32,185]. When p-type silicon wafers (0.3 mm thick, ρ_{Si} = 25
Ω·cm) are used, the resulting optical structures are found to have

ϵ = 0.16 at room temperature. At 100°C, the estimated normal blackness of these structures is ϵ_n = 0.15, whereas the corresponding value for structures based on silicon with ρ_{Si} = 1 $\Omega \cdot$cm is 0.3.

These measurements have thus shown that the deposition of selective multilayer coatings on high-efficiency solar cells that are transparent in the infrared will increase α_S to 0.9 (because of the absorption of solar radiation in a broader spectral range) and, at the same time, will reduce the thermal emission coefficient to $\epsilon \leqslant 0.16$ because of the high infrared reflection coefficient of the structure. The electrical output characteristics of solar cells are not affected by this, but the high values of the ratio α_S/ϵ ensure that the thermal component of the efficiency of the photothermal system or collector is higher.

Optimization of solar cells for photothermal installations using concentrated radiation

The parameters of solar cells made from a semiconducting material with high initial efficiency under extra-atmospheric or terrestrial conditions are found to deteriorate sharply with increasing concentration if their series resistance is 0.5-0.6 $\Omega \cdot$cm^2, which is typical for most solar cells used in conventional batteries. Even for relatively low concentration ratios (C = 20-50), for which thermal and photoelectric solar energy converters with selective coatings [23] are particularly efficient, the loss of efficiency and electric power generated by the solar cells may be so high that composite photothermal systems and collectors with concentrators may become energetically ineffective and expensive. The initial electrical performance and efficiency of high-grade solar cells can be maintained at higher concentration ratios by depositing on the thin, uppermost doped layer (which has the greatest influence on the series resistance) a dense contact grid with optimized strip dimensions and semitransparent metal layers (with antireflective properties on one or both sides), which sharply reduce the surface resistivity of the doped region.

The parameters of solar cells designed for operation under enhanced radiation intensity were optimized for cells with a two-component doped layer structure (see Fig. 2.1) in which the impu-

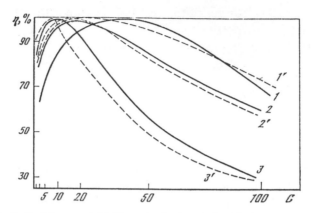

FIG. 3.10. Measured (1-3) and calculated (1'-3') efficiency of solar cells as a function of concentration ratio (AMO conditions) for different values of the series resistance: 1, 1') 0.1; 2, 2') 0.2; 3, 3') 0.5 $\Omega \cdot cm^2$.

rity concentration in region I is constant while, in region II next to the p—n junction, the exponential distribution of impurities corresponds to a constant built-in electric field [304]. The jump in potential across the separation boundary between regions I and II is $\ln(N_1/N_2)KT/q$.

This model provides an adequate description of the measured impurity distribution in the doped layer produced in silicon by modern diffusion techniques, especially in the case of diffusion through an anode film with controlled porosity [122,125]. It was assumed that the base region was uniformly doped. The dependence of carrier mobility and diffusion length on the concentration of impurities was described by the relation $\mu \sim N^{-0.5}$. The collection coefficient was calculated from formulas similar to those given in [122]. The only difference was that a large number of cell parameters [304] was taken into account, e.g., rate of surface recombination and carrier concentration in the base.

Both possible ways of reducing the series resistance R_s of cells (by depositing a dense contact grid and multilayer coatings, including a semitransparent metal layer) were used to produce special solar cells for comparison with theoretical calculations and experimental data (Fig. 3.10). Good agreement between them was observed. When $R_s \cong 0.1$ $\Omega \cdot cm^2$, the maximum efficiency is attained for concentration ratios of 20-50.

AM2 measurements showed that the efficiency of these cells for concentration ratios of 30-70 was 12-13%.

Our studies have thus led to the development of silicon solar cells with optimized electrical and optical parameters and the necessary performance in photothermal devices with high concentration ratios.

3.3   Selective coatings for the free surfaces of photothermal systems and thermal collectors

Solar cells do not occupy the entire working surface of the composite photothermal system, and the ratio of the area free from cells to that covered by them depends on which particular component of the converted solar energy is to be obtained in the form of electric power and which in the form of thermal power. The free surface of the composite system is best covered by a selective coating with a high α_S/ϵ ratio in order to achieve an additional increase in the efficiency of conversion of solar radiation into thermal power. These coatings are not subject to any stringent restriction (as compared with coatings for solar cells). All that is necessary is to ensure that the ratio α_S/ϵ is high and the coating is highly transparent in the spectral sensitivity range of the solar cells. Like the coatings for solar collectors [23,218], these coatings can be opaque to all wavelengths in the solar spectrum because, when the coating thickness is small, the thermal resistance (to the transmission of heat from the surface of the coating to the heat-receiving elements of the installation or collector) is very small and has practically no effect on the thermophysical parameters of the system as a whole.

Coatings for photothermal installations that are not covered by solar cells, and for thermal collectors, can be divided in accordance with their application and basic working conditions: (1) coatings for installations and collectors with planar configuration and an air gap between the outer insulating glass and the surface absorbing solar radiation and (2) coatings for evacuated (usually tubular) photothermal and thermal collectors.

Coatings in the first subgroup must have $\alpha_S \geqslant 0.9\text{-}0.92$ (at moderate working temperatures of the order of 50-60°C, the coefficient α_S has a greater effect than ϵ on the efficiency of conversion of solar radiation into power), but can have the relatively low ratio α_S/ϵ = 3-5 because radiation losses constitute only a fraction of the overall thermal loss from flat-plate collectors. The presence of the air gap between the glass and the absorbing surface, which contains water vapor, means that the coatings must be unaffected by high humidity levels in the presence of local hot-spots and intensive insolation. They must also have exceptional resistance to damage by heat, light, and corrosion. Since aluminum and its alloys are the most common materials for flat-plate collectors, and are not resistant to the effects of humidity in the absence of special shielding or protective cover (for example, plating), the deposition of stable selective coatings will extend the overall life of solar-power installations.

Selective coatings in the second group will work under near-ideal conditions: the heat-receiving surface is completely shielded from unfavorable atmospheric effects by placing it in an evacuated enclosure, and the thermal balance of evacuated tubular collectors is almost entirely determined by the values of α_S and α_S/ϵ. The question of the maximum possible ratio α_S/ϵ is then of both theoretical and practical interest. Coatings with a very high value of α_S/ϵ for evacuated tubular collectors have been optimized by calculation and then produced experimentally. These studies have also led to the maximum attainable values of α_S/ϵ for the selective black-and-white coatings (insofar as is possible with current thin-film technology and the use of substrates with maximum reflection in the infrared).

Selective optical coatings for flat-plate collectors

Selective multilayer coatings deposited by evaporation in high vacuum [309,310]. The thermal and light stability of these coatings can be substantially enhanced by using layers of high-melting dielectric oxides, such as zirconium dioxide, and transparent films of high-melting metals, such as chromium [311]. These structures consist of successive stable layers of zirconium dioxide + chromium + zirconium dioxide + chromium and are best deposited by electron-beam evaporation in a high vacuum onto thin aluminum, steel, nickel, or copper foils held at 300-400°C, which can then be attached to the surface of the flat-plate collector with a thermally conducting and elastic organosilicone rubber sealant or adhesive [309]. This method has a number of technological advantages and can be used to produce selective coatings on large-area collectors without preliminary polishing of the surface. Coatings have been obtained with α_S = 0.88-0.92, ϵ = 0.11-0.12 (measured at 30°C), and stable properties up to 250-350°C.

Selective coatings for large-area flat-plate collectors, produced by the traditional electrochemical deposition of black nickel and chromium. The thermal and light stability of coatings of this type is relatively poor, but they continue to attract the attention of researchers [218,312].

Selective coatings produced by thermal decomposition of organometallic compounds. Selective coatings based on black films of Co_3O_4 were prepared by thermal decomposition of cobalt acetylacetonate, $(C_5H_7O_2)_3Co$, evaporated at 150-200°C onto aluminum foils heated to 350-450°C [313]. When the thickness of the Co_3O_4 layer was 0.3 μm, the integrated optical coefficients α and ϵ were found to be 0.95 and 0.08, respectively (at 30°C), so that $\alpha_S/\epsilon \cong$ $\cong$ 11.9. These coatings survived prolonged tests in high-humidity chambers and exposure to simulated solar radiation for 1440 hr.

Similar coatings, with comparable values of α_S and ϵ, were deposited by pulverization of a mixture consisting of half molar solution of cobalt nitrate and thiourea in the ratio of between 1:1 and 1:1.66. The rate of pulverization by compressed air was 2 ml/min [314]. When the black cobalt films were deposited on an aluminum foil to a thickness of 0.21 μm, it was found that α_S = = 0.91-0.92 and ϵ = 0.13 (both measured at 100°C). The foil dimensions were 20 × 20 cm and the foil itself was held at between 130 and 180°C during the deposition process. The coatings were found to retain their properties when heated up to 220°C [314].

Surface of collector with selective optical properties due to given surface roughness. When the polished collector plates are subjected to the mechanical grinding process, the linear dimensions of the resulting regular surface pits become comparable with the wavelength of solar radiation. However, the plates continue to behave as reflecting mirrors for long-wave infrared radiation [315].

Selectively rough surfaces can be produced by exploiting the properties of dendritic films produced by high-temperature chemical deposition of, for example, metallic rhenium from the vapor phase.

The sawtooth dendritic structure of the surface of the rhenium film enhances the ratio α_S/ϵ of the absorbing surface of solar collectors [312].

A simpler method of producing selective coatings with the dendritic structure is as follows. A stainless steel foil, 50 μm thick, containing chromium (4%) and aluminum (0.3%), is heated for a few seconds at 900°C in an argon atmosphere containing 0.1% oxygen. This produces a very thin imperfect oxide film on the surface of the foil, which is essential for the appearance of the dendritic oxide layer produced by subsequent oxidation in air for 8 hr at 900°C. The length and width of the resulting dendrites (whiskers) were found to be 2.5 μm and 0.25 μm, respectively. The separation between the dendrites was 0.5-1.0 μm. The length-to-width ratio of the dendrites can be increased to 20:1 by modifying the oxidation process. This substantially improves the selective optical properties of oxide coatings on steel [316].

Selective coatings based on low-vacuum metal condensates. The high absorbing power of these coatings is due to the low values of the reflection coefficient at most solar wavelengths, and the complete absorption of incident radiation in the interior of the coating because of the large number of pores and the highly developed specific area. The coatings consist of metal particles with a relatively uniform distribution in the host medium. The latter is a mixture of oxides of the same metal, formed during the condensation of the film in the relatively poor vacuum. However, such coatings usually have poor chemical and mechanical stability [23]. Selective coatings based on low-vacuum aluminum condensates, with improved corrosion and thermal parameters, have been produced by improving the deposition process and using controlled baking [317,318]. Stable coatings have been produced by simultaneous condensation from a supersaturated aluminum vapor and deposition from a molecular beam. Oxidation of the finely dispersed condensate particles was also found to improve their mutual bonding and the adhesion of the coating to the substrate. Coatings of this kind have typically α_S = 0.9-0.98 and ϵ = 0.12-0.15. If necessary, ϵ can be raised to 0.8 by increasing the thickness of the coating and modifying the deposition parameters. This means that this procedure can be used to produce stable nonselective coatings for receivers of infrared radiation.

Two-layer electrochemical coating on aluminum surface. A thin film produced by anodic oxidation is deposited on the surface of the metal. The film pores adjacent to the surface are filled (by electrochemical implantation) with light-absorbing particles of a metal such as nickel. The outer layer of the aluminum dioxide film does not then contain nickel atoms and exhibits antireflective properties [319]. When the electrochemical process takes place, it is exceedingly difficult to achieve uniform anodization to a thickness not exceeding a fraction of a micrometer (or 1-2 μm) over the entire surface area of the aluminum collector, amounting to 0.7-1.2 m^2 (this requires a uniform current density distribution over the entire surface to be anodized). This is an important point because an increase in the thickness of the film produced by anodization would give rise to a sharp increase in the emission coefficient ϵ. It is well known that, when the thickness of the aluminum dioxide film exceeds 10 μm, the emission coefficient ϵ reaches 0.78-0.8.

A three-layer electrochemical coating consisting of aluminum and its alloys, which is free from this drawback, has been suggested [320,321]. The first layer is produced directly on the surface of the aluminum collector and consists of an anodic aluminum dioxide film with pores filled with light-absorbing metal particles. The second layer consists of hydrated aluminum dioxide (which is necessary to improve the adhesion between the first and third layers of the coating), and the third layer is a transparent, conducting film of tin dioxide. The coating as a whole has a low emission coefficient (ϵ = 0.18-0.25) because of the high infrared reflecting power of the uppermost tin dioxide layer, so that ϵ does not depend on the thickness of the first layer or on the preliminary treatment of the collector surface. The optimized layer thicknesses producing the best combination of optical, mechanical, and working parameters of the three-layer coating are as follows: first layer — 2-5 μm, second layer — 4-10 μm, third (outermost) layer — 0.4-0.6 μm [320]. The coefficient α_S of this coating lies in the range 0.9-0.94.

Collectors made from the AD-1 aluminum alloy were anodized in an 18% water solution of sulfuric acid for 40 min in a current of 1.5 A/dm^2, the electrolyte temperature being 18-20°C. The anodized specimens were then carefully washed with water and electrolytically colored by passing an alternating current through a solution of nickel, copper, and tin salts. An electrolyte of the following composition was used to implant copper atoms in the film: $CuSO_4 \cdot 5H_2O$ — 35 g/l; $MgSO_4 \cdot 7H_2O$ — 20 g/l; H_2SO_4 — 5 g/l; the pH of the electrolyte was 1.2-1.4. The whole procedure was performed in a bath for 5 min at 15 V and then for a further 5 min at 20 V [321]. The emission coefficient ϵ of the collector surface consisting of the AD-1 aluminum alloy, anodized and colored with copper salts, was 0.87. The collector surface was then heated to 360-400°C and the outermost transparent conducting tin dioxide layer was deposited by the aerosol method ($SnCl_4$ was sprayed on in air from a gun at 2-4 atm for 20-30 min). The final value of ϵ was 0.18-0.25 (at 30°C).

Of course, to simplify the technological process of fabrication of these selective coatings, all the layers can be produced in the same way. An attempt was therefore made to develop a fabrication procedure in which each layer in the coating was deposited by hydrolysis during the chemical pulverization process.

Two-layer selective coatings produced by chemical pulverization. Currently available technology can produce these coatings on any metal collector surface, heated to about 400°C [322]. The first layer is black CuO, deposited by evaporating a water solution of $Cu(NO_3)_2$. The second layer is a transparent conducting tin dioxide film produced by pulverization of tin chloride. The time taken to deposit the two-layer coating does not exceed a few minutes but, to reduce internal stresses, coatings consisting of several tens of very thin films have been made [287], each film (50-500 Å) being deposited from a specially developed pulsed pulverizer in a period of 1—2 sec.

Two-layer coatings with integrated coefficients α_S = 0.88-0.91 and ϵ = 0.25-0.29 have been produced, and field tests have begun with collectors under normal operating conditions. No changes in the optical parameters of the coatings were detected after heating to 350-400°C for 100 hr. The deposition of an antireflective film of silicon dioxide on top of the tin dioxide film, using chemical pulverization and hydrolysis, was found to raise the value of α_S to 0.92-0.94 [320-322].

Selective optical coatings for evacuated tubular collectors

Coatings with high values of α_S/ϵ are desirable in the case of evacuated tubular collectors, but resistance to corrosion is then of minor importance.

The use of such coatings allows radiation damage to solar batteries to be cured by annealing, and materials to be heated in vacuum by solar radiation [323]. For coatings with high α_S/ϵ, the temperature of the metal plates or tubes inside an evacuated transparent envelope can be several hundred degrees centigrade even without solar concentrators.

The following are the most promising methods of producing selective coatings for evacuated tubular collectors:

(a) vacuum deposition of alternate dielectric and semitransparent metal films;

(b) electrochemical deposition of black layers; and

(c) simultaneous evaporation in vacuum of a dielectric and a metal (or a preprepared mixture thereof) with a view to producing metal-ceramic films.

Before experimental tests can begin, we have to solve the problem of optimization of selective coatings in order to estimate the maximum values of α_S/ϵ that can be obtained with the above methods of producing such coatings.

The M-4030 computer was used to calculate the optical parameters of multilayered structures [324,325]. For all the models of selective coatings considered, the reflection coefficient $R(\lambda)$ was computed in the wavelength range 0.3-50 μm, using the methods described in [23,292-295]. The computed functions $R(\lambda)$ were then used to calculate α_S under AM0 and AM2 conditions, the normal emission coefficient ϵ_n at 27-500°C, and the corresponding ratios α_S/ϵ_n. The deformed wavelength scale described in [23,46] was employed. Coating parameters such as type of material and layer thickness were optimized with respect to the ratio α_S/ϵ_n. In experimental tests, measurements were also made of the hemispherical emission coefficient ϵ_h. The calculated data can be corrected with the aid of the ratio $\epsilon_h/\epsilon_n = 1.25\text{-}1.30$ [46].

Selective vacuum coatings. Substrate materials that were considered included silver, aluminum, copper, and nickel. The optical constants of thick (of the order of 0.1 μm) metal films were taken from [46] for aluminum and silver, from [46,294,326] for copper, and from [326,327] for nickel. Silver and copper have the lowest and nickel the highest values of α_S and ϵ_n. This determines the sequence in which the metal layers are deposited in multilayer systems. The preferred materials for substrates are silver, copper, and aluminum; nickel is suitable for intermediate layers.

Calculations have been carried out of the optical parameters of systems consisting of one-, two-, three-, and four-layer thin-film coatings comprising not only transparent dielectric layers, but also combinations of transparent and partially absorbing thin metal films. The following were selected out of a large number of dielectrics: silicon dioxide (refractive index n = 1.45), aluminum dioxide (n = 1.7), and zinc sulfide (n = 2.3). It is well known that the optical constants of thin (50-200 Å) metal films are rapidly varying functions of thickness. Their refractive index usually increases with decreasing thickness and the films themselves acquire semiconducting properties [293,294]. However, published data consist of optical constants of thin metal films for certain values of the thickness and wavelength only, or they are given in a very narrow spectral range, so that the optical constants of thick condensed metal films had to be used in the calculations.

These calculations have shown that transparent dielectric coatings (even multilayer coatings) do not ensure sufficient antireflective properties in the solar wavelength range. The coatings must include partially absorbing layers (for example, nickel), which means that a dielectric with the maximum possible refractive index has to be used to ensure high enough α_S. The material chosen was zinc sulfide (n = 2.3), which has better infrared transparency than, for example, silicon dioxide.

Table 3.3 lists the maximum values of α_S/ϵ_n (AM2 conditions, 100°C) that can be achieved with a four-layer coating of the form nickel + zinc sulfide + nickel + zinc sulfide on silver, copper, or aluminum substrates.

Electrochemical coatings. The dependence of the refractive index and absorption coefficient of black nickel films on the current density used in electrochemical deposition [328] can be exploited to produce coatings of, for example, successive films of "interference" thickness with high and low refractive indices and a variety of overall optical properties. The layered structure of this type of coating can be obtained by varying the current density in accordance

TABLE 3.3
Calculated Maximum Ratio α_S/ϵ_n for Four-Layer
Coatings on a Metal Substrate

Substrate	Optimum layer thickness, Å				α_S/ϵ_n (AM2, 100° C)	
	Ni	ZnS	Ni	ZnS	without coating	with coating
Ag	50	200	50	300	2.065	44.599
Cu	50	200	50	300	11.088	38.352
Al	150	200	50	300	4.923	27.648

with a predetermined program (an increase in the current density produces a reduction in the refractive index). It is also possible to achieve a continuous variation in the chemical composition and structure, as well as a gradual reduction in the refractive index and absorption coefficient with distance from the surface of the substrate.

Substrate materials that have been considered include copper and nickel, which can be first electrolytically deposited on the polished aluminum or steel collector surface [324,325,329]. Nickel has a higher initial degree of blackness than copper, but it is interesting to estimate the optical parameters of selective coatings deposited on a nickel substrate because it resists corrosion, and coatings deposited on nickel should have greater stability. Moreover, nickel can have a lower reflection coefficient in the solar range and, hence, a higher absorption coefficient α_S, which is important when coatings for nonevacuated thermal collectors are produced.

The optical constants n and k of black nickel films [328] were assumed constant in the wide spectral range 0.3–20 μm. In the case of single-layer coatings, the highest values of α_S/ϵ_n can be obtained for n = 1.77 and k = 0.43, which corresponds to the highest admissible current densities used in the deposition process. The maximum α_S/ϵ_n at 100°C is achieved for a black nickel layer with l = 900 Å and is equal to 13.31 under AM2 conditions on a nickel substrate and 56.94 on a copper substrate. The corresponding AM0 figures are somewhat lower: α_S/ϵ_n = 12.88 and 54.81, respectively.

For two-layer electrochemical coatings, the optimum combina-

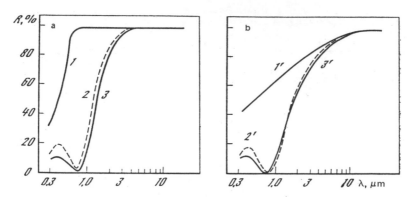

FIG. 3.11. Wavelength dependence of the reflection coefficient of the polished surface of copper (a) and nickel (b) covered with an electrochemical coating of black nickel: 1, 1') before deposition of antireflective layer; 2, 2') single-layer coating ($n_1 = 1.77$, $k_1 = 0.43$, $l_1 = 900$ Å); 3, 3') two-layer coating ($n_1 = 1.77$, $k_1 = 0.43$, $l_1 = 600$-700 Å; $n_2 = 2.07$, $k_2 = 0.98$, $l_2 = 200$-300 Å).

tion of optical constants of black nickel ($n_1 = 1.77$, $k_1 = 0.43$, and $n_2 = 2.07$, $k_2 = 0.98$) is obtained when the outermost layer has the lowest and the innermost the highest refractive index. The maximum α_S/ϵ_n of such coatings is however, only slightly higher than for optimum single-layer coatings: for the two-layer coatings on nickel substrate at 100°C under AM2 conditions, it is found that $\alpha_S/\epsilon_n = 13.34$ ($l_1 = 700$ Å, $l_2 = 200$ Å), whereas for the two-layer coating on copper substrate $\alpha_S/\epsilon_n = 58.61$ ($l_1 = 600$ Å, $l_2 = 300$ Å). Optimization of three-layer electrochemical coatings has revealed that they are no more effective than the two-layer coatings.

Figure 3.11 shows the calculated reflection coefficient of one- and two-layer electrochemical coatings with optimized parameters on nickel and copper substrates.

Metal-ceramic coatings. Similar calculations of the optical parameters have been performed for structures consisting of cermet films produced by the simultaneous evaporation in a high vacuum of a metal and a dielectric from two sources or from a tablet prepared by compacting a powdered mixture of the metal and dielectric (in which case, an electron beam or thermal evaporator was employed). It was assumed that $n = 1.77$ throughout the wavelength range 0.3-20 μm, whereas $k = 0.43$ for $\lambda \leqslant 3.0$ μm and $k = 0.10$ or

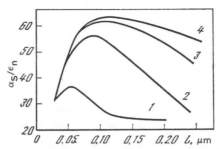

FIG. 3.12. Calculated α_S/ϵ_n under AM2 illumination, at collector temperature of 100°C, as a function of the thickness of selective coatings: 1) ZnS—Ni—Cu; 2) black nickel—Cu; 3, 4) cermet film—Cu for different refractive indices (k = 0.10 and 0.05, respectively).

0.05 for $\lambda > 3.0$ μm. These values of the optical constants are very similar to those published for cermet films consisting of mixtures of nickel and magnesium oxide or nickel and silicon dioxide [330].

Figure 3.12 shows the calculated α_S/ϵ_n as a function of the coating thickness on a copper substrate. For the vacuum coating (curve 1) with the zinc sulfide + nickel + copper composition, the maximum α_S/ϵ_n are shown for different values of the total thickness l of the coating (for l = 500 Å, the thickness of the nickel layer was $l_2 \cong 130$ Å whereas for l = 1200 Å it was $l_2 \cong 170$ Å).

More complicated vacuum coatings, for example, zinc sulfide + + nickel + zinc sulfide + nickel + copper, have only slightly higher α_S/ϵ_n (38.4 instead of 37.3).

In contrast to the electrochemical black nickel coating (curve 2), cermet films have a much lower absorption coefficient for $\lambda > > 3$: k = 0.10 (curve 3) and k = 0.05 (curve 4), which produces higher α_S/ϵ_n, especially for the higher values of the thickness.

By varying the electrochemical deposition conditions, it is possible to produce black nickel coatings with a more favorable wavelength dependence of n and k, similar to that obtained for cermet coatings [321].

The above calculations lead to the following two important practical conclusions: (1) the above methods can be used to produce selective coatings with high values of α_S/ϵ_n and (2) electrochemical and metal-ceramic coatings consisting of fine and uniformly distributed absorbing metal particles in a transparent dielectric medium have greater optical selectivity and higher α_S/ϵ (for the same total coating thickness) than vacuum coatings consisting of successive layers of a dielectric and a semitransparent metal.

Experiment has confirmed these calculations. It was found that $\alpha_S > 0.9$ could be obtained for all the above types of coating, but it was only for the cermet and electrochemical coatings that the emission coefficient of the substrate did not change and remained, for example, at 0.02-0.03 in the case of the polished copper collector which suggests that there is a real possibility of reaching $\alpha_S/\epsilon_n = 30$-40 for the absorbing surfaces of evacuated tubular collectors. Electrochemical coatings in the form of two-layer black nickel structures [329] have been shown to have sufficient stability under illumination and heating, provided they are deposited from baths containing nickel and zinc chlorides and not the usual sulfate electrolytes.

Optimum interval of concentration and working temperature for collectors with selective optical coatings

By introducing the idea of the perfect selective surface and the optimum threshold wavelength (at which low reflection in the main part of the solar spectrum is rapidly and conveniently replaced with high reflection in the range corresponding to the intrinsic thermal emission of the surface), we were able to calculate the optimum range of solar concentration and working temperature for collectors with selective optical coatings, and determine the extent to which the parameters of practical selective coatings differed from the limiting theoretical values. These calculations are usually performed [23] under AM0 conditions. The influence of the selective optical properties of the collector surface on its efficiency under AM0 and AM2 conditions was estimated in [331]. This is of considerable practical importance for the most extensive application of thermal and photothermal collectors, i.e., domestic heating, air-conditioning, hot water supplies, and the generation of electric and thermal power for solar houses.

We must now examine in greater detail the value of selective optical coatings and establish the range of the concentration factors

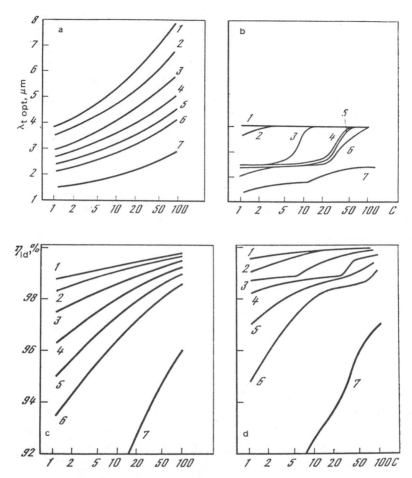

FIG. 3.13. Optimum threshold wavelength (a, b) and thermal efficiency (c, d) of the ideal selective surface as functions of concentration C under AM0 illumination (a, c) and under AM2 illumination (b, d) for different working temperatures: 1-7) 70, 100, 150, 200, 250, 300, and 500°C, respectively.

for which a selective surface of given temperature has a higher thermal efficiency than the surface of a perfect black body.

In the ideal case, the absorbing power of a selective surface is a step function, the threshold wavelength λ_t being defined by the following conditions: $\alpha(\lambda) = \epsilon(\lambda) = 1$ for $\lambda < \lambda_t$ and $\alpha(\lambda) = \epsilon(\lambda) = 0$ for $\lambda > \lambda_t$. It is clear that the optimum value $\lambda_{t\ opt}$ for which the absorbing surface retains the maximum amount of useful energy depends both on the concentration ratio and on the converter temperature.

Figure 3.13 shows the calculated $\lambda_{t\ opt}$ and maximum thermal efficiency of the ideal selective surface as functions of the concentration ratio C for different working temperatures. These functions were obtained from the heat balance equation, written in the most general form. Under radiative heat transfer conditions, the heat absorbed is given by

$$Q_a = C\sigma_S E - \varepsilon\sigma T^4,$$

where σ is the Stefan-Boltzmann constant and E is the incident solar flux density.

The maximum efficiency of the ideal selective surface is then given by

$$\eta_{id} = \alpha_S - \varepsilon\sigma T^4 / CE.$$

Under terrestrial conditions,

$$\eta_{id} = \alpha_S - \varepsilon\sigma (T^4 - T_0^4)/CE,$$

where the average seasonal temperature of the ambient air is $T_0 = 10°C$ (it is assumed that heat is mostly lost by reradiation, which is valid for evacuated converters). The calculations were based on the tabulated intensity distributions in AM0 [332], AM2 [333], and black-body spectra in the wavelength range between 0 and λ [46].

Since the solar spectrum extends into the region $\lambda > \lambda_{t\ opt}$, where $\alpha(\lambda) = 0$, and part of the black-body emission lies in the range $\lambda < \lambda_{t\ opt}$, where $\epsilon(\lambda) = 1$, it follows that even the ideal selector surface has an efficiency less than 100%. However, at temperatures below 250°C, which are typical for composite photothermal collectors, the thermal conversion efficiency of the ideal selective surface under AM0 illumination will exceed 95%. This figure rises to 97% under AM2 illumination (see Fig. 3.13c and d). The complex nature of the optimum threshold wavelength and the efficiency of the ideal surface as functions of concentration under the AM2 spectrum is explained by the presence of absorption bands

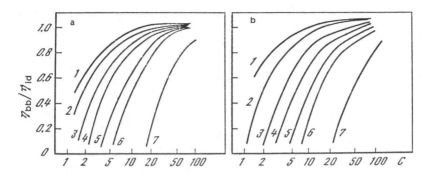

FIG. 3.14. Ratio of the thermal efficiency of the surface of a black body (η_{bb}) to the efficiency of the ideal selective surface (η_{id}) as a function of the concentration factor C under AM0 (a) and AM2 (b) illumination at different working temperatures: 1-7) 70, 100, 150, 200, 250, 300, and 500°C, respectively.

after the radiation passes through the Earth's atmosphere. A similar result has been obtained for the AM1.5 spectrum.

The energy conversion efficiency of the black-body surface, η_{bb}, and of the ideal selective surface, η_{id}, can be compared by using the graphs given in Fig. 3.14. For given C, the necessity for the selective surface increases with increasing temperature. At given temperature, the selective surface is more effective when the solar concentration is low. The black-body surface illuminated by highly concentrated light can reradiate only the single energy flux, so that η_{bb} approaches η_{id}.

Analysis of these results shows that, for T < 300°C and C = = 1-20, selective surfaces produce higher energy conversion efficiency than the black body. Moreover, selective surfaces have a great advantage over black surfaces under AM2 conditions as compared with AM0. In solar energy converters working on the ground without concentrators and using water as the coolant (working temperature 70-100°C), the optically nonselective surface at T = 100°C (Fig. 3.14b, curve 2) will not generate useful thermal energy but, for C = 10, its efficiency amounts to 90% of the efficiency of the ideal selective surface. When it is necessary to produce a working temperature of 200°C (Fig. 3.14b, curve 4), the black-body surface begins to produce useful heat only for C > 3.3, and the efficiency of the black body for C = 10 amounts to 68% of the efficiency of the ideal surface. When T = 300°C, the two surfaces have comparable efficiencies for C ⩾ 100 and, finally, when T ⩾ 500°C, the black-body efficiency approaches the efficiency of the ideal surface for concentration ratios much greater than 100.

The above data on $\lambda_{t\ opt}$ can be used to choose the selective surface with the optimum optical characteristics for the particular converter working conditions.

It is obvious that selective optical coatings must be used in composite photothermal converters whose working temperature is restricted because of the temperature dependence of the efficiency of solar cells for the most readily attainable concentration ratios C = 1-20. The physical characteristics of solar cells in composite photothermal converters have therefore been optimized mostly for this range of concentration ratios.

It is important to note that, when an optical coating has an appreciable selectivity within the solar wavelength range (for example, it has appreciable interference maxima and minima), the replacement of AM0 with AM2 (or AM1.5) in the calculation of α_s has an appreciable effect on this quantity because the absorption bands of water vapor and ozone, and also aerosol scattering, influence the ratio of the infrared to visible components of the spectrum. The terrestrial solar spectrum must therefore be used in calculations of α_s for selective coatings [334].

Determination of the emission coefficients of solar collectors before and after the deposition of selective coatings

Before we can calculate the thermal balance in solar energy converters and estimate the efficiency of solar installations, we must have detailed knowledge of the integrated thermal emission coefficient ϵ of coatings on the solar collectors at the most commonly used temperatures (60-90°C). However, most of the devices currently used in solar technology to monitor ϵ operate at room or very high temperatures (in excess of 1000°C) [23,335-337].

A thermoelectric sensor (TES) has been used to measure the thermal emission coefficient of surfaces used in solar power engineering at the working temperatures, before and after the deposition of selective optical coatings [338]. The principle of this device is

TABLE 3.4

Measured Integrated Emission Coefficient of Selective Coatings
for Solar Installations and Collectors

Substrate + coating	$\varepsilon_{30°\,C}$ (FM-63)	$\varepsilon_{90°\,C}$ (TES)
Polished aluminum + multilayer interference coating of the form $Ni-SiO_2-Ni-SiO_2$ (deposited by evaporation in vacuum; $l_{Ni} = 0.01\text{-}0.015\ \mu m$; $l_{SiO_2} = 0.08\text{-}0.09\ \mu m$) [310]	0.09	0,085
Aluminum foil + multilayer coating of the form $Ni-SiO_2-Ni-SiO_2-Ni-SiO_2$ ($l_{Ni} = 0.01\text{-}0.015\ \mu m$; $l_{SiO_2} = 0.08\text{-}0.09\ \mu m$) [310]	0.13 0.16 0.18 0.18 0.16 0.13	0.138 0.162 0.183 0.135 0.162 0.135
Polished aluminum + black selective coating based on low-vacuum aluminum condensates [317,318]	0.16	0.162
Copper foil + nickel-base coating (evaporated in a vacuum, $l_{Ni} = 0.2\text{-}0.3\ \mu m$) [309]	0.12 0.11 0.12 0.11	0,129 0.103 0.121 0,104
AD1 aluminum + oxide film (electrochemical anodization [320,321])	0.4	0.395
Polished aluminum + oxide layer with implanted nickel atoms (anodization and electrochemical blackening by implantation of nickel atoms; $l_{Al_2O_3} = 2\text{-}3\ \mu m$) [320,321]	0.32 0.27 0.32	0.347 0,307 0.328
Polished aluminum + black acrylic enamel (AK-512)		
$\quad l = 1 - 2\ \mu m$ $\quad l = 10 - 15\ \mu m$	0.56 0.58 0.8 0.9 0.85 0.85	0.546 0.595 0.778 0.92 0.835 0,835
Electrochemically polished aluminum + thick oxide layer (anodization; $l_{Al_2O_3} = 18\text{-}20\ \mu m$)	0.7	0.753

that it measures ϵ by comparing the radiation from a given surface with that from a perfect black body. A sensitive radiometer incorporating the TES is described in [103,105].

The device was used at 90°C to measure the integrated emission coefficient $\epsilon_{90°C}$ of different surfaces used in solar power engineering. The results were then compared with $\epsilon_{30°C}$ obtained with the conventional FM-63 thermoradiometer at 30°C (Table 3.4).

It is important to note that, whenever tests indicate that a given selective coating is thermally stable, the integrated thermal emission coefficient ϵ at different temperatures can be calculated from the wavelength dependence of the infrared reflection coefficient of the coated collector. Such calculations are based on nomograms using a distorted wavelength scale [23,46]. The scales shown in Fig. 3.15 were calculated from tabulated black-body spectra at different temperatures.

These results have been used to determine the temperature dependence of the integrated emission coefficient ϵ of transparent conducting coatings made from tin oxide, indium oxide, and their mixtures. These coatings retain their spectral characteristics up to 700-800°C, especially when they are deposited on glass or quartz. The following values were calculated for two transparent conducting coatings consisting of 10% tin dioxide and 90% indium trioxide, i.e., ITO coatings (the reflection curves are shown in Fig. 3.6, curves 1 and 2) with different surface resistivity [0.02 kΩ/$\square$(ϵ_1) and 0.1 kΩ/$\square$(ϵ_2)] :

t, °C	27	277	427	527
ϵ_1	0.165	0.191	0.209	0.235
ϵ_2	0.275	0.334	0.374	0.413

The precision of these calculated values of ϵ is, of course, largely determined by the precision of the measured spectral reflection coefficients of the selective coatings. Unfortunately, only the specular component of the reflection coefficient can be measured at present with existing infrared spectrophotometers. Whenever the diffuse component of the reflection coefficient is small (ITO coatings), results similar to those given above provide a reasonable measure of the temperature dependence of the integrated emission coefficient.

3.4   Tests on selective optical coatings for solar cells, thermal and composite photothermal systems, and collectors used on the ground and in space

Selective optical coatings enhance the efficiency of conversion of solar energy into thermal and electric power, and modify the

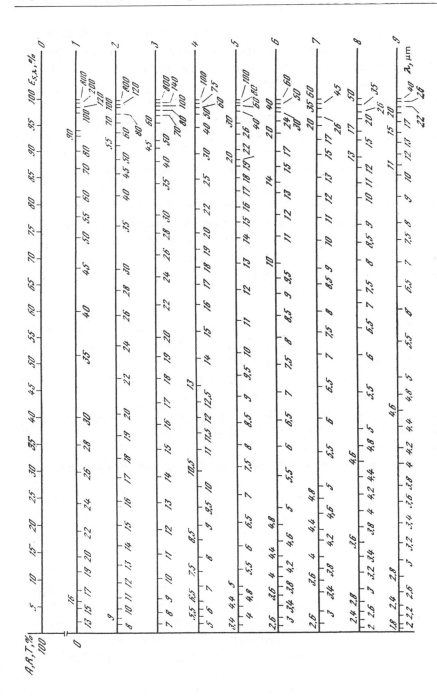

FIG. 3.15. Isoenergy (0) and deformed (black body) wavelength scales for different temperatures: 1-9) −153, −93, −53, −27, 147, 277, 327, 427, 527°C, respectivey.

working temperature in the required direction. They ensure that the amount of absorbed radiation increases as a result of the anti-reflective properties of the surface; they protect the surface from corpuscular radiation, and they improve the stability of conversion characteristics under prolonged operation.

Composite photothermal systems, which produce both thermal and electric power, can be compared qualitatively by considering the overall efficiency, written as a simple sum of electrical and thermal efficiencies. However, it must be remembered that, whenever the efficiency is determined and compared for different devices, different forms of energy must be reduced to a common denominator, and the parameters influencing the operation of the system as a whole must be the same. The overall efficiency of a photothermal collector can be correctly determined by assuming that the thermal energy generated by it is used to produce electric power at the operating temperature of the solar cells. It is only then that the overall efficiency of the device can be calculated by summing the energy generated by the solar cells and the thermal collector in the form of electric power. This simplified approach to the calculation of the efficiency of composite systems is justified in many of the cases considered in this chapter only because the efficiency is determined only to enable us to compare the quality of selective optical coatings and not to optimize the mechanical, electrophysical, and other characteristics of energy converters.

Flat-plate thermal collectors with selective optical coatings

Since, for flat-plate themal collectors with an air gap between the glass and the absorbing surface, a substantial fraction of the thermal loss is due to convection through the thermally insulating layer, the use of selective coatings will result in an increase in the working temperature of the coolant used in the collector by no more than 10-20°C. The efficacy of using selective coatings is then very dependent on the temperature of the working surface of the collector, and is found to increase substantially as the coolant flow rate is reduced. One- or two-layer electrochemical coatings based on black chromium and nickel are usually deposited on collectors of this type [339]. Stable interference-type coatings [340] consisting of nickel + cerium dioxide + silicon dioxide, nickel + silicon monoxide + magnesium fluoride, or nickel + zinc sulfide + magnesium fluoride films [309,311] have been successfully tested. The use of these coatings with α_S = 0.9-0.91 and ϵ = 0.05-0.06 (an ordinary nonselective surface has α_S = ϵ = 0.94) has resulted in an increase by a factor of 1.5—2 in the productivity of collectors used to heat water to 70°C.

Evacuated tubular thermal collectors with multilayer selective coatings

Evacuated tubular collectors were investigated over a number of years (1975-1979) and various improvements were achieved. This has resulted in a gradual increase in their efficiency from 30% to 67%. Multilayer selective coatings (incorporating successive layers of a dielectric and a semitransparent metal) are now employed on these collectors and produce α_S/ϵ = 19 [341]. Improved thermal insulation of the pipes between the evacuated units, and the use of additional concentrating reflectors made from aluminum foil and placed between these units, have raised the efficiency of the active heat-receiving surface of the collector to 78-82%.

Composite photothermal sytems and collectors with solar cells and selective coatings

Early designs of composite photothermal systems, incorporating single-crystal or thin-film solar cells located on the surface of a conventional flat-plate thermal collector [302,303], had an efficiency of no more than 50% (conversion into electric power — 3%, conversion into thermal power — 47%). The solar cells in these systems were protected (as in the batteries used in space) by various types of glassing, which increased the integrated emission coefficient to 0.86-0.9, but also produced a rise in thermal losses into the ambient medium.

Designs of photothermal converters with higher overall efficiency are described in [106,342,343]. Convection losses were almost totally eliminated by using an evacuated tubular enclosure, while the use of selective coatings with a low emission coefficient ϵ on the surface of the solar cells (placed on the inner tube carrying the coolant), or on the surface of the glass envelope facing the cells,

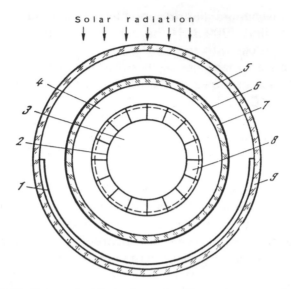

FIG. 3.16. Composite photothermal collector incorporating solar cells: 1) reflecting metal film; 2) heat-absorbing surface; 3) tube containing liquid or gaseous coolant; 4) space filled with transparent silicone liquid; 5) evacuated space; 6) transparent selective coating with low ϵ; 7, 9) transparent glass enclosures; 8) single-crystal or thin-film solar cells (flat or tubular).

has resulted in a sharp reduction in losses by radiation [307,308]. The composite photothermal collector with two tubular glass enclosures (Fig. 3.16) is particularly interesting. It enables the solar cells to be inserted into the transparent, inert, organosilicone liquid [106, 342]. The presence of the transparent liquid gives the inner envelope some focusing properties, thereby increasing by a factor of 1.2-1.3 the amount of solar energy reaching the solar cells. Moreover, the higher thermal capacity ensures that the collector acquires heat-storage properties, and the life of the collector is much greater because of the protection afforded by the inert liquid which stabilizes the properties of the outer and end surfaces of the solar cells. Tests have shown that the overall efficiency of composite photothermal collectors of this design is in excess of 70%, of which 10-12% is due to the electric energy generated by the silicon solar cells [106, 343]. The amount of electric power in the overall power output of the composite photothermal collector can be raised by 20-22% by replacing the silicon solar cells with AlGaAs—GaAs cells.

Terrestrial solar cells and batteries in sealed
protective shells

There are two ways of protecting terrestrial solar cells and batteries, their antireflective coatings of "interference" thickness, and their contacts from atmospheric effects during operation under unfavorable climatic conditions:

(1) external protective glass or polymer with glass fiber reinforcement [344,345] can be attached with a layer of the radiation-stable silicone rubber or a fluoroelastomer (another possibility is to triplex thin solar cells between two layers of glass or a fluoroelastomer film, using hot compression molding);

(2) solar cells assembled in modules and groups can be inserted into gas-filled tubular glass enclosures [346].

The latter method will stabilize the parameters of thin-film copper sulfide—cadmium sulfide solar cells which are sensitive to external influences [347]. The stability of the properties of thin-film solar cells can be substantially improved by adjusting the composition of the gas mixture in the glass shell and by depositing a semitransparent copper layer (of the order of 100 Å) on top of the copper sulfide, followed by thermal treatment, the deposition of antireflective dielectric coatings, and the deposition of transparent lacquer.

Tests on modules of thin-film solar cells based on the copper sulfide—cadmium sulfide heterosystem (the copper sulfide layer was deposited from a solution of cuprous chloride) were performed over a period of four years. The modules were inserted into sealed, gas-filled containers and the tests were performed under field conditions near Moscow. They showed that, when optimum protective coatings were employed and the working volume was filled with dry gas, the degradation effect did not exceed 5% over the test period.

Tests on transparent heat-reflecting and conducting
coatings

The optimum combination of high transparency to solar radiation and low surface resistivity and, consequently, low integrated emission coefficient ϵ, can be achieved with the following two types of selective coating:

(1) the three-layer titanium dioxide + silver + titanium dioxide structure [46] or the zinc sulfide + silver + zinc sulfide structure [23,307,308] deposited on glass by thermal evaporation in a high vacuum;

(2) transparent conducting oxides based on doped wide-gap semiconductors (tin dioxide, indium trioxide, or their mixtures; cadmium stannates) deposited on glass, usually by chemical pulverization during the hydrolysis in air of sprayed solutions of metal salts, or by plasma-ion sputtering in vacuum at a low inert-gas pressure.

Three-layer heat-reflecting semitransparent silver coatings on glass, with antireflective zinc sulfide layers on both sides (zinc sulfide + silver + zinc sulfide system) with integrated solar transmission coefficients $T_S = 0.65$ and thermal emission coefficients $\epsilon = 0.06$, have been used in vacuum experiments on the annealing of radiation defects (produced by corpuscular radiation in the Earth's radiation belts) in solar battery modules. The experiments are illustrated in Fig. 3.17. One minute after the AMO Sun simulator producing $E = 1360$ W/m^2 on the surface of the solar battery module inserted into the vacuum chamber was turned on, the temperature of the system was found to be 320°C without any other source of heat, and this was sufficient to anneal the radiation defects. The integrated thermal emission coefficient of the module was 0.9 on the front surface and 0.95 at the rear. When the transparency of the heat-reflecting coating through which the simulated solar radiation entered the module was increased, the equilibrium annealing temperature could be raised to 380-400°C [323].

Transparent conducting coatings based on mixtures of indium and tin oxides (ITO films) have been used to shield the system from the electric component of the electromagnetic field produced by solar batteries themselves [301]. This shielding was used on the Intercosmos—Bulgaria-1300 satellite, which carried scientific equipment for continuing composite studies of physical processes in the Earth's iono- and magnetosphere. The satellite was launched on 7 August 1981 [348]. The shielding was also used on the Aureole-3 satellite carrying scientific equipment developed by Soviet and French specialists as part of a Franco-Soviet project (Arcad-3). This

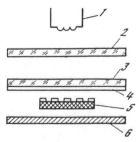

FIG. 3.17. Experiment on vacuum annealing of radiation defects in a silicon solar battery module: 1) AMO simulator; 2) light filter; 3) glass; 4) transparent heat-reflecting coating of ZnS—Ag—ZnS; 5) silicon solar battery module; 6) heat-reflecting screen of aluminum foil with $\epsilon = 0.04$.

was designed to investigate processes occurring in the Earth's magneto- and ionosphere as well as the nature of polar auroras. The satellite was launched into orbit on 21 September 1981 [349].

The solar batteries on both satellites had electrical connections running in opposite directions and were provided with radiation-shielding glassing on both sides of the modules. The outer surfaces were coated with transparent, conducting ITO coatings (surface resistivity of the order of 1 kΩ/$\square$). None of these arrangements, including the electrical connections within the battery circuit and to the body of the satellite [350,301], produced any detectable effect on the operation of the scientific equipment on board the satellite. Experiments performed by the Soviet-Bulgarian and Soviet-French Earth satellites showed that, in addition to the variety of problems solved successfully by using optical coatings on solar batteries, it has also been possible to solve the substantial scientific problem of "electromagnetically clean" batteries. Similar studies were performed earlier with the Explorer-31 and Geos satellites, on which the batteries were screened from the electric field of the internal electromagnetic environment by transparent electrically conducting coatings of indium oxide [21]. By optimizing the thickness and resistivity of the conducting films, their antireflective, radiation-shielding, and emissive properties could be held at the previous level.

Studies of enhanced radiation-shielding properties
of charge storing glass

The Cosmos-605, -690, and -782 satellites launched in 1973-1975 carried biological objects and were used to perform experiments on radiation hazards in space. They included studies of new and promising methods of radiation shielding (from the effects of charged particles in space) and shielding based on the ability of the electrostatic field in the interior of a dielectric to deflect charged-particle beams. The first of these space experiments were concerned with the possibility of maintaining a strong field in the interior of a dielectric in which the field was produced by a high-voltage generator on board the satellite. In subsequent missions, the deflecting field was produced by trapping electrons encountered by the spacecraft equipped with dielectric shields as it passed through the Earth's radiation belt [351].

The experimental results obtained by studying the stability of the charged state and the optical properties of glasses used as radiation shields for solar cells and batteries on the above satellites in space and on Cosmos-936 (1977) are reported in [352].

Four special containers mounted on the exterior of the spacecraft were used to test radiation shielding and optical properties of coatings in space. Each container was provided with a lid. Sample coatings were inserted into a container and held on a special plate divided into compartments. The oriented motion of the satellite was maintained only during ascent. During its flight in orbit, the satellite was not stabilized and rotated slowly, so that the effect of radiation (both corpuscular and electromagnetic) was the same in all containers. During the ascent, the container lids were opened under the heat shield. They were closed again during the descent, and were returned to the Earth together with the satellite. The sample coatings were exposed for 20 days on each satellite. The mean height of the satellite orbit above the Earth's surface did not exceed 300 km and the inclination was about 62°.

The glass specimens selected for investigation were very similar to phosphate glass [284,285]. Terrestrial tests showed that they were capable of supporting strong internal electric fields under electron bombardment, and could maintain these fields for a considerable period of time after the bombardment process. Before they were mounted in the above containers, the glass specimens were irradiated in the LUE-8-5 linear electron accelerator (electron energy 6.2 MeV) and the KGU-300 electrostatic accelerator (0.2 MeV). This established the necessary internal electric field. The beam current in the accelerator and the electron flux were the same for all the glass specimens (0.5-5 μA and 10^{14}-10^{15} cm^{-2}, respectively).

The thickness of the specimens inserted into the two accelerators was greater than the ionization range of the incident electrons. It was 0.7 and 0.1 of the glass thickness, in the two cases, respectively, i.e., the relaxation of both the volume and surface fields was investigated. Control specimens for comparative tests were kept on the ground.

The following measurements were performed after the specimens were returned to the Earth (at the end of the mission occupying τ months): (a) stimulation by impact was used to establish the presence of the electric field in the specimens, (b) the γ-ray probe method [353] was used to determine the potential distribution at a depth of about 1 mm on each side and the radiation dose on the specimens in orbit, and (c) the SF-4 spectrophotometer was used to measure the optical transmission of the glasses before and after the space mission.

The temperature of the internal surfaces of the container and of the glass specimens in orbit did not exceed 60°C, and the corpuscular radiation dose absorbed by the glasses was 0.14-0.2 Gy. Low ion fluxes with different atomic numbers and different energies were also recorded on the glass surfaces during the flight of Cosmos-690 and -782. These results are in agreement with the data obtained with Skylab.

The γ-ray probe method of measuring the electric potential φ in glass relies on the emission of Compton electrons and photoelectrons from the specimen surface when it is exposed to Co^{60} γ-rays. The thickness of the probed layer was about 1 mm on either side of the specimen. The change in the emission of these electrons by a charged dielectric when the space charge is removed (by heating at 250°C for 4 hr) is related to the initial layer potential (Table 3.5)

The spread in the values of the potential is due to the difference between the electron beam currents in the two accelerators used to irradiate the specimens. There is practically no difference between the potentials determined for control specimens and specimens exposed in space.

TABLE 3.5
Space-Charge Stored in Glass Specimens Recovered from Space

Satellite	Specimen	τ, months	Charge present	τ, months	φ, kV
Cosmos-690	Charged	8	Yes	5	10—110
	Uncharged	8	No	5	—
	Control	8	Yes	5	10—180
Cosmos-782	Charged	4	Yes	3	80—380
	Uncharged	4	No	3	—
	Control	4	Yes	3	80—480
Cosmos-936	Charged	7.5	Yes	5	7—20
	Uncharged	7.5	No	5	—
	Control	7.5	Yes	5	4—30

Measurements of the optical transmission of phosphate glass before and after the mission of the Cosmos-782 satellite showed that there was practically no change in the high initial transparency of the glass. In the long-wave part of the spectrum (0.6-0.7 μm), there was a small increase in transparency, probably due to the anti-reflective film that appeared on the glass surface during the degassing of organic adhesives and binders in space (these materials were used to attach the glasses to the internal compartments in the container).

It is important to note that, when initially uncharged specimens were examined after return from space, there was no detectable electric field in their interior, which is explained by the relatively low orbits used in these experiments (well under the Earth's radiation belts).

The experiments showed that the charged state of dielectrics in low orbits in space (and on the ground) was characterized by low rates of relaxation and persisted for long periods of time.

When glassed solar cells are exposed to 1-MeV electrons in an accelerator producing a beam flux of 10^{16} cm^{-2}, the electric power generated by the cells after irradiation (for the same initial power) is higher by 25-30% when charge-storing glasses are used (as compared with ordinary radiation-shielding glass).

It may therefore be concluded that both terrestrial and space experiments have confirmed that the shielding properties of optical coatings on solar cells are enhanced by the interaction between electrons and the electric field in a dielectric, when strong enough fields are produced as the spacecraft crosses the Earth's radiation belts.

Silicon solar cells and batteries transparent in the infrared and equipped with heat-reflecting mirror coatings

Silicon solar cells transparent in the infrared and provided with heat-reflecting mirror coatings [109-111,141] have a low equilibrium working temperature and can be widely used not only in batteries for automatic interplanetary stations, such as those in the Venera series [142] which operate under high levels of illumination, but also for studying direct solar radiation as well as the radiation reflected from the Earth and its cloud cover. This has been confirmed by field tests reported in [354].

Solar batteries mounted on low-orbit satellites (200-400 km) must be protected from overheating due to the Earth's albedo and its thermal emission. The Earth's albedo and, consequently, the thermal regime of solar batteries, are very dependent on the cloud cover and the optical properties of the underlying surface in the particular locality. The amount of reflected solar radiation reaching the back surface of solar battery panels may, in turn, be determined by solar-cell probes facing the Earth (and shielded from direct solar light), provided their temperature in orbit can be determined by some independent method.

All these interrelated questions were investigated in experiments performed on Cosmos satellites, in particular, Cosmos-1061, -1280, and -1301. These satellites were designed to investigate the natural resources of the Earth in the interests of the Soviet economy and international cooperation.

Two rectangular panels bearing small solar batteries and sensors (Fig. 3.18) were mounted inside the container carrying the scientific equipment of these satellites. The lower panel could be accurately oriented so that its normal pointed toward the center of the Earth, whereas the upper panel was at right angles to the direction of incidence of solar radiation at any time in orbit. Four small solar batteries were mounted on each panel to ensure the reliability and precision of the final results. Each battery consisted of 78 flat solar-cell modules of 54 × 40 mm and was similar to the solar

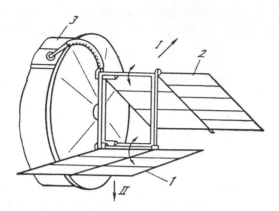

FIG. 3.18. Disposition of flat panels carrying small solar batteries on some of the Cosmos satellites: 1, 2) panels facing the Earth and Sun, respectively; 3) container housing the scientific equipment; I) direction of the Sun; II) direction of the Earth.

batteries used on the automatic interplanetary stations Venera-9 and Venera-10 [142]. The solar cells and the supporting panel were transparent to the solar infrared in the wavelength range between 1.1 and 2.5 μm, and were shielded from cosmic rays and the solar ultraviolet by thin radiation-shielding glasses attached to the antireflective surface of the silicon solar cells by transparent silicone rubber (the optical properties of the glasses and the rubber were found to be stable in the course of prolonged tests in space [347]). Heat-reflecting aluminum coatings were deposited on the inner face of the shielding glasses (above contact junctions between the cells) [141]. Glasses attached to the rear of the modules were also coated with a heat-shielding aluminum layer (see Fig. 2.12) which reflected not only rays with wavelengths between 1.1 and 2.5 μm, transmitted by the transparent solar cells, but also 83-85% of the solar spectrum (0.3-25 μm) incident on the rear surface after reflection from the Earth and its cloud cover.

Two batteries on each panel were operating in the short-circuit current mode, and the current was used to determine the amount of solar radiation incident on the Earth (panel 2) and reflected from the Earth (panel 1). The open-circuit voltage of the other batteries was used to determine the panel temperature (the temperature

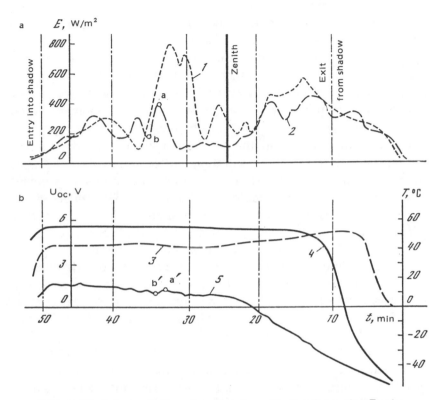

FIG. 3.19. Solar radiation flux density reflected from the Earth and from its cloud cover, measured by one of the Cosmos satellites in orbit (a), and variation in the open-circuit voltage and temperature of solar batteries (b): 1, 2) first and second orbits, respectively; 3) open-circuit voltage generated by batteries facing the Sun; 4, 5) temperature of batteries facing the Sun and the Earth, respectively.

dependence of this voltage was determined in laboratory experiments).

Panel 2 also carried a special Sun detector consisting of solar cells [17], whereas panel 1 carried an electromechanical ampere-hour sensor, which was reset every time the reading of 10 A·hr was reached.

All the small solar batteries used in these experiments had very similar short-circuit currents, carefully measured under AM0 Sun simulators in the laboratory. An initial reference point was thus obtained (the short-circuit current of batteries under 1360 W/m^2), and was then used to determine the flux density of radiation reflected from the Earth at any time in orbit.

Figure 3.19 shows the time distribution of solar radiation reflected from the Earth, which is typical for low-lying orbits. These data were obtained by one of the Cosmos satellites in 1981 in an orbit with minimum duration of illuminated segment (about 52 min). This type of orbit requires a higher output power generated by the solar batteries, and the radiation reflected by the Earth is the "reserve" necessary for generating additional power [143,149]. The orbital planes characterized by minimum illuminated segments lie in the plane of incidence of the light flux. The reflected flux density usually rises on either side of the zenith line because there is then a sharp improvement in reflection by the cloud cover and in the incidence of reflected flux on the back of the panels.

The maximum temperature of the panels is low (it does not exceed 57-58°C in contrast to the 70-75°C, typical for solar cells of conventional design [110]). This applies even to batteries on panel 2 (Fig. 3.18) because the cells are transparent to the solar infrared and heat-reflecting coatings are employed. It is clear from Fig. 3.19 that the temperature of panel 1 (Fig. 3.18) is a function of the amount of radiation incident upon it after reflection from the Earth (points a′ and b′ on curve 4 correspond to points a and b on curve 2).

The above studies have demonstrated that the working temperature of solar cells and batteries can be substantially reduced by using heat-reflecting mirror coatings that are transparent to the solar infrared.

4

Determination of the Efficiency and Metrological Parameters of Solar Cells and Batteries

It is well known that, before the efficiency of solar cells and batteries (or, indeed, the efficiency of any other radiation converters) can be determined, we must know the amount of radiant energy received by the solar cell and the amount of electric power generated by it in a given time. However, the problem is complicated by a number of factors:

(1) the energy arrives in the form of solar radiation, whose spectral composition and power are still being investigated (even under AM0 conditions), while the parameters of terrestrial solar radiation are very dependent on the state of the atmosphere and frequently vary over very short periods of time;

(2) the development of Sun simulators that reproduce most of the AM0 parameters, or the parameters of terrestrial solar radiation, gives rise to complex scientific and technological problems;

(3) when stable reference solar cells are developed for Sun simulators, it is essential to take into account the optical and electrical properties of each type of cell, including their spectral sensitivity; when the output electrical parameters of solar cells and batteries are measured, it must be borne in mind that the series resistance and the resistance of the measuring devices have a considerable influence on the final results.

It is thus clear that the determination of the efficiency of solar cells and batteries is a complex problem. The metrology of semiconductor solar energy converters has in fact become a separate topic in photoelectricity.

Metrological problems, and questions relating to the precise determination of the efficiency of solar cells, are intimately related to studies of their optical characteristics. We recall that the spectral and integrating optical instruments and methods are quite general, the surface properties of solar cells have a decisive effect on their efficiency, the solar spectrum must be accurately simulated in measurements of both the efficiency and optical parameters of cells, and the optical coatings on the working surfaces of solar energy converters and components of Sun simulators must satisfy stringent conditions. All this provides a close link between two major and important branches of solar power engineering.

The development and standardization of accurate methods of measuring the characteristics of solar cells and batteries have attracted considerable attention in all countries involved in the utilization and conversion of solar radiation into other forms of energy. General interest in this field has been stimulated by considerable advances in the performance of silicon solar cells [2,5,13, 17-22] and gallium arsenide heterojunction cells [1,9,115-117, 156-168]. This means that solar batteries can already be widely used not only in space, but also under terrestrial conditions, and we now have a basis for developing photoelectric power stations capable of delivering appreciable amounts of power [1-3]. Problems relating to the stability of the parameters of modern solar cells and batteries under prolonged operation can now be regarded as essentially solved [13,142,178,183,191,347].

The electrical parameters of solar cells have to be measured for a number of reasons. For example, they are essential when standardizing parameters have to be determined in quality control, when technological processes have to be monitored, when efficiency sorting is carried out prior to selection of individual cells and groups for assembly with minimum switching losses [13,21], when the electrical parameters have to be predicted for different nonstationary conditions of operation, and in the optimization of parameters when new types of solar cell are developed [5,142,149,354]. The demands on the conditions of measurement and on experimental precision may be very different in different cases. One of the most important problems is the determination of the parameters of a batch of manufactured cells prior to certification. Such measurements must be strictly standardized, and accurate enough, so that reliable and reproducible results can be obtained. Standardization of measurements is essential for the accurate prediction and determination of the characteristics of batteries when they are designed, developed, and fabricated.

4.1 Solar radiation and the choice of the standard spectrum

The problem of accurate measurement is intimately related to the precise reproduction of standard parameters of solar radiation such as flux density, spectral and angular distribution of energy, and flux uniformity and stability.

Extra-atmospheric radiation

When the parameters of solar cells for use in space are determined, the standard radiation is always taken to be the solar radiation intercepted on a plane perpendicular to the direction of the Sun at a distance equal to one astronomical unit from it (this is the average Earth–Sun separation). The energy received from the Sun per unit area per second under these conditions is called the *solar constant*. The angular diameter of the Sun subtended at this distance is 31'59'' [355], so that each point on an illuminated elementary area receives a beam of rays within a cone of angle ±16'. The radiation flux is perfectly uniform.

The spectral energy distribution of solar radiation has frequently been measured, both on the ground and outside the atmosphere. However, values of the solar constant measured under different conditions do not agree. The US Committee on Solar Radiation has adopted as standard the data obtained by M. P. Thekaekara and A. J. Drummond, who took an average over extra-tropospheric measurements [356,357]. The adopted value of the solar constant was 1353 W/m^2. E. A. Makarova and A. V. Kharitonov have derived the spectral distribution by taking an average over all reliable measurements, both terrestrial and high-altitude [358]. When measurements performed outside the atmosphere alone are considered, the solar constant turns out to be 1360 W/m^2. C. W. Allen [359] has taken an average over the spectral distribution, taking into account virtually only terrestrial measurements, and again found that the solar constant was 1360 W/m^2. The same measurements were used as a basis for the model of solar radiation described in [360].

Analysis of available information on the parameters of solar radiation [361] indicates that the spectral distribution put forward by Makarova and Kharitonov is preferable. Their distribution is used by the European Space Center [362].

Attempts are continuing to obtain more accurate values for the solar constant. Measurements from spacecraft and rockets have resulted in a weighted average of 1370 W/m^2 [363]. The variation in the solar constant due to the periodic variation in solar activity has frequently been investigated [364]. Analysis of terrestrial measurements of the solar constant [365] shows that its root mean square deviation due to phenomena occurring on the Sun amounts to ±0.1%, while possible oscillations in absorption within the Earth's orbit account for ±0.14%. High-altitude measurements have shown that, during the second half of the 22-year solar cycle, the solar constant changes by not more than 0.75% [363]. Further studies using stabilized spacecraft should provide data on long-term variations in the solar constant.

The actual conditions under which space cells are used are not very different from the conditions adopted as standard. The spectral energy distribution of solar radiation (averaged over the disk) is constant over the entire region of space in which spacecraft operate. The angular divergence of the beam does not vary too much: it amounts to about ±42', ±22', ±11', and ±3' at the distance of the orbit of Mercury, Venus, Mars, and Jupiter, respectively. Assuming that the solar constant was 1360 W/m^2, the solar flux density at the top of the Earth's atmosphere in 1980 was found to deviate from the mean within ±3.5%, i.e., from 1406 W/m^2 at the beginning of January of each year (when the Earth is at the minimum distance from the Sun) to 1315 W/m^2 in July, when the Earth is at the most distant point of its orbit (Table 4.1) [366].

More detailed information on solar radiation outside the atmosphere is summarized in the review given in [367].

When two-sided solar batteries [146,5] or batteries transparent to the solar infrared [109-111] are designed for low-orbit Earth satellites [143,149], it is important to take into account the Earth's albedo for the incident solar radiation [365,368].

Limb darkening has often been reported, and the thermal emission of the Earth and other planets, as well as their albedos, all of

TABLE 4.1
Solar Flux Densities on Planetary Orbits

Planet	Solar flux density, W/m²		
	Mean distance from Sun	Perihelion	Aphelion
Mercury	9071	14388	6242
Venus	2599	2634	2565
Earth	1360	1406	1315
Mars	586	713	490
Jupiter	50.2	55.5	45.7

which are necessary for thermal calculations involving spacecraft and solar batteries, are available [367].

The measured AMO solar spectrum has been found to differ from the spectrum of the perfect black body held at 5785 K (this is the most common approximation). The apparent brightness at the center of the solar disk is greater by a factor of 1.22 than the average brightness. The brightness of the disk increases toward the limb, and there is also a variation in the emission spectrum (the red part of the spectrum becomes stronger toward the limb) because the color temperature at the limb is lower than that at the center.

The value of the solar constant has therefore had to be modified several times in the course of the last fifty years. The value used in the early work on solar cells [82] in 1923 was 1350 W/m² (proposed by C. G. Abbot). In 1954, F. S. Johnson [369] obtained 1393 W/m², while the standard adopted at the beginning of the 1970s was 1353 W/m², as deduced by M. P. Thekaekara [356,357]. The value regarded as the most reliable at present is that derived by E. A. Makarova and A. V. Kharitonov, i.e., 1360 W/m² [356-362,366,370].

Once we know the absolute value of the solar constant, we can determine the energy reaching the surface of a solar cell or battery working outside the atmosphere. However, to determine the useful electric power generated by a solar cell, we must also perform an accurate measurement of the spectral distribution of the incident radiation, especially within the spectral sensitivity range of modern solar cells (in the case of silicon cells, 0.3 to 1.1 μm).

It turns out that, within the relatively narrow spectral range 0.3-1.1 μm, the total amounts of radiation incident on silicon solar cells, obtained by using the solar constants reported throughout the literature, are not very different [370] : 991 W/m² [356] ; 1039 W/m² [369] ; 1014 W/m² [358].

If we compare the different spectral energy distribution curves obtained under AMO conditions, we find that between the solar emission maxima and within the spectral sensitivity range of silicon solar cells (0.6-0.8 μm), the distribution obtained by Johnson is closer (despite the very different solar constant) to the distribution of Makarova and Kharitonov than that of Thekaekara.

This conclusion was confirmed when the integrated photocurrent of silicon solar cells was determined from the spectral distribution of solar radiation (using the spectral dependence of solar cell sensitivity) and when field measurements made on Malta were extrapolated to zero air mass [370]. It we take the photocurrent calculated using the Johnson distribution as being 100%, the integrated photocurrent determined from the Makarova-Kharitonov spectrum turns out to be 99.3%, whereas that determined from the Thekaekara spectrum is 95.7%, i.e., very different from the other two.

The experiment performed on Malta [370] is in excellent agreement with calculations based on the spectrum given in [358].

The AMO spectral distribution of solar radiation put forward by Makarova and Kharitonov [358] is the best distribution to use in determinations of the efficiency of all semiconductor solar cells and batteries.

Terrestrial solar radiation

The choice of the standard parameters for terrestrial radiation is complicated by the considerable variation in the conditions under which terrestrial solar cells have to operate. The intensity and spectrum of solar radiation on the Earth's surface depends on the position of the Sun above the horizon, the height of the particular locality above sea level, the state of the atmosphere, and the optical properties of the underlying surface. The altitude of the Sun above the horizon determines the path length in the atmosphere. A special quantity, called the *optical air mass* m, is usually introduced. Unit air mass corresponds to the path traversed by solar rays when they are incident vertically at sea level. For a plane-parallel model of the atmosphere, the optical air mass at sea level is practically equal to the cosec of the altitude of the Sun above the horizon. For the actual atmosphere, this relation is well satisfied from $10°$ upward [371]. Air masses (at sea level) of 1, 1.5, 2, 3, and 5 correspond to the following altitudes of the Sun above the horizon: $90°$, $41°49'$, $30°$, $19°27'$, and $11°32'$, respectively. The air mass is also a function of the height of the locality above sea level: it decreases with this height in proportion to the atmospheric pressure. The air mass is zero at the top of the atmosphere.

The air mass is assumed to be equal to unity at sea level under clear, cloudless sky, when the Sun is in the zenith and its radiation is incident normally on the surface of the measuring cells (the atmospheric pressure is then $p_0 = 1.013 \cdot 10^5$ Pa).

The air mass at any point on the Earth's surface can be determined from the equation

$$m = p/p_0 \sin \theta = p \cosec \theta/p_0, \qquad (4.1)$$

where p, θ are, respectively, the pressure and angle defining the altitude of the Sun above the line of the horizon at a given point on the Earth's surface ($p_0 = 1.013 \cdot 10^5$ Pa).

The composition of the atmosphere has a substantial effect on the radiation parameters. Radiation is absorbed and scattered as it passes through the atmosphere. Absorption is due to the components of the atmosphere, including water vapor, ozone, oxygen, carbon dioxide, and so on, but is largely determined by water vapor. Scattering is due to gas molecules (Rayleigh scattering) and aerosols. Aerosol scattering depends on the amount and size of dust particles suspended in the atmosphere.

When Rayleigh scattering is taken into account, the transmission of the atmosphere can be estimated from the following formula [372], which gives the proportion of solar radiation transmitted by the atmosphere after Rayleigh scattering:

$$\tau_r = \exp\left(-0,008735\lambda^{-4,08}\, mp/p_0\right).$$

Transmission reduced by absorption by water vapor is characterized by the fraction of solar radiation transmitted by the atmosphere within the spectral region corresponding to the absorption bands of water:

$$\tau_\omega = \exp(-k_\omega(\lambda)\,\omega),$$

where $k_\omega(\lambda)$ is the absorption coefficient of water vapor for solar radiation and ω represents the layer of precipitated atmospheric water vapor.

It is important to note that absorption by water vapor and the permanent components of the atmosphere, such as ozone, oxygen, carbon dioxide, and ammonia gas, is very selective. Although there are empirical relationships for calculating the absorption due to each of these atmospheric components, Fig. 4.1 gives a much clearer idea of what happens to solar radiation on its way to the Earth [373].

Aerosol scattering can be estimated in terms of the "turbid atmosphere." The direct solar flux attenuated by aerosol scattering can be determined from the following formula [374]:

$$\tau_a = \exp(-\beta\lambda^{-\alpha}m), \tag{4.2}$$

where β is the turbidity coefficient and α is the selectivity index [375].

The turbidity coefficient represents the amount of particles suspended in air, while the selectivity index represents the size distribution of the particles: the smaller the particles, the smaller the index α, and the greater the fraction of radiation attenuated in the ultraviolet and blue parts of the spectrum. It is assumed that, depending on atmospheric conditions, α lies in the range 0.8-2.0 and β in the range 0.01-0.375.

When the generalizing formula that takes into account all types of loss of solar radiation as it passes through the terrestrial atmosphere was derived [376], it was assumed that the relation between the solar spectral flux density E_λ within a narrow wavelength interval and the spectral density $E_{0\lambda}$ outside the atmosphere within the same interval was

$$E_\lambda = E_{0\lambda}\exp\left(-(c_1+c_2+c_3)m\right)T_{\lambda i}, \tag{4.3}$$

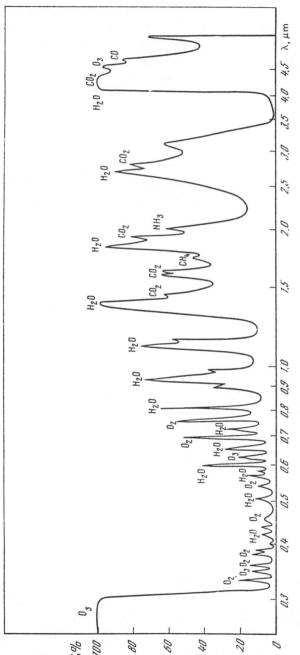

FIG. 4.1. Selective spectral absorption of solar radiation by the permanent components of the atmosphere under air mass m = 1, thickness of precipitated water vapor equal to 2 cm, and reduced ozone layer thickness equal to 2 mm (normal temperature and pressure).

where c_1, c_2 and $c_3 = \beta\lambda^{-\alpha}$ are, respectively, the optical path length corresponding to Rayleigh scattering, the ozone layer, and the dust content of the atmosphere, and $T_{\lambda i}$ is a factor representing the reduction in atmospheric transparency due to the molecular absorption bands, which can be written (depending on the position of each band) as follows:

$$T_{\lambda_1}=\exp(-c_4(\omega m)^{1/2}), \qquad T_{\lambda_2}=\exp(-c_5\omega m), \qquad T_{\lambda_3}=1-c_6 m^{1/2},$$

where c_4-c_6 are empirical constants [377,378].

Various models of the atmosphere have now been developed and can be used to compute the optical transmission of the atmosphere for solar radiation [379].

The terrestrial spectra of direct solar radiation for air mass between 0 and 5 and constant parameters of the atmosphere ($p_0 = 1.013\cdot10^5$ Pa, $\omega = 2$ cm, reduced thickness of ozone layer 2.8 mm, number of dust particles in air $d \approx 300$ cm^{-3}) have been calculated (Fig. 4.2) from the AM0 spectrum [82] using the formula

$$E_\lambda=E_{\lambda 0}\exp(-\alpha_\lambda m), \qquad\qquad (4.4)$$

where α_λ is the absorption coefficient due to the individual components of the atmosphere in a narrow spectral range [380]. The transmission of the atmosphere with allowance for the aerosol scattering was calculated not from (4.1) but from the formula

$$\tau_a=\exp(-1{,}02\cdot10^{-4\lambda-0{,}75d}).$$

We can use these terrestrial solar spectra and other calculated and experimental spectra (see, for example, [381]) to estimate the efficiency of solar cells made from different semiconducting materials under different climatic and geographic conditions. It must be remembered, however, that solar cells working without concentrators convert not only the direct but also the diffuse solar radiation, including the component due to molecular Rayleigh and aerosol scattering by the atmosphere. The diffuse component of sky radiation may be very considerable even on clear days (Fig. 4.3) [374,382]. Experimental data on the resultant and diffuse solar fluxes for $m = 1$ are given in [383].

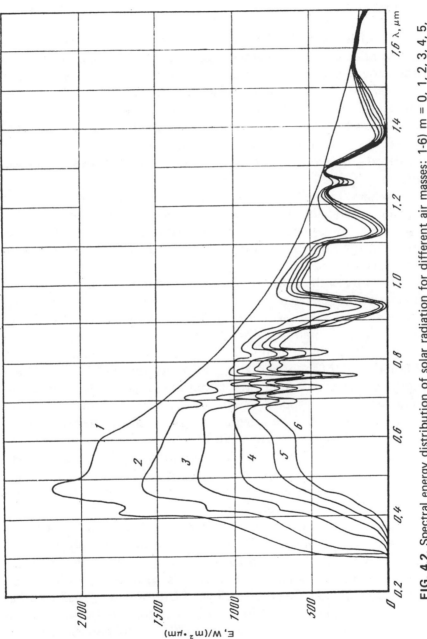

FIG. 4.2. Spectral energy distribution of solar radiation for different air masses: 1-6) m = 0, 1, 2, 3, 4, 5, respectively.

International standardization of the terrestrial
solar spectrum

The efficiency of solar cells produced in different laboratories and manufactured commercially cannot be compared without introducing some standard methods for estimating their output parameters. It is particularly important to use standard methods to measure the characteristics of solar cells and batteries working under terrestrial conditions because the electric power generated by selectively sensitive solar cells is not uniquely related to the flux density of terrestrial solar radiation which has a variable spectrum.

Standardization of the methods of measurement on both the national and the international scale will facilitate collaboration in solar energy utilization, and will help to resolve problems in comparative evaluation of solar cells and batteries manufactured in different countries.

The Fifteenth Session of the International Commission on Illumination (CIE), held in Vienna in 1963 and attended by representatives from Great Britain, USSR, USA, and other countries, recommended a standard for the simulation of terrestrial solar radiation. It defined standard illumination conditions as those corresponding to a horizontal plane under air mass m = I (AM1 conditions) and the following atmospheric parameters: layer of precipitated water vapor − 2 cm, ozone layer − 2 mm, turbidity coefficient β = 0.05. The integrated flux density of terrestrial solar radiation was then adopted as 1110 W/m².

Questions relating to standard conditions of illumination were examined by CIE in subsequent years. In 1972, CIE recommended that the value of the solar constant used in simulations of conditions in space be taken to be 1350 W/m² ± 5% [384].

For aging tests on materials under illumination, and for calculations involving the energy of terrestrial solar radiation, the Commission recommended that Moon's data [380] on the spectral distribution of solar radiation on the Earth's surface for different air masses be employed. The proceedings of CIE should be read together with the review given in [385] of different computational formulas [386] and models of the atmosphere [379] put forward by different authors.

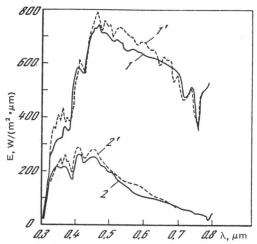

FIG. 4.3. Spectral energy distribution of the total
(1, 1′) and diffuse (2, 2′) terrestrial solar radiation
for m = 2 and β = 0.1: 1, 2) calculated; 1′, 2′) experimental.

Many researchers have published data on different aspects of
solar radiation [385] and these are useful in calculations of the
parameters of terrestrial solar cells and batteries. They include the
spectral dependence of the direct ($\gamma = E_d/E$) and scattered ($\gamma' = E_s/E$) components of total solar radiation E for different positions
of the Sun above the horizon (Fig. 4.4). The relative contribution
of scattered radiation increases not only as the Sun becomes lower
above the horizon, but also with decreasing wavelength, as is clearly
indicated by Fig. 4.4. The relative amount of scattered radiation is
particularly large in the ultraviolet and at the blue end of the visible
spectrum for all positions of the Sun above the horizon.

Interesting measurements of the spectral composition and
intensity of solar radiation transmitted by the cloudy atmosphere
are also reported in [385]. If the solar flux density E on a cloudless
day is taken to be 100%, then a 20% cloud cover (0.2 of the sky
covered by clouds) reduces E to 89%, while covers of 40, 60, and
80% reduce E to 77, 64, and 46%, respectively. Under a continuous
cloud cover, the solar flux density falls to 20%. The correlated color
temperature of the Sun, as seen from the Earth on a cloudy day, is
6020-6050 K.

Active work began in various countries in 1974-1975 on the

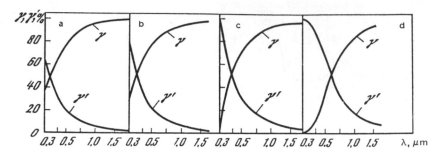

FIG. 4.4. Relative amounts of direct (γ) and scattered (γ') radiation for different altitudes of the Sun above the horizon: a-d) 90, 60, 30, and 10°, respectively.

definition of the standard spectrum of terrestrial solar radiation that could be used in measurements of the parameters of solar cells and batteries. A standard solar spectrum corresponding to air mass m = 1 was put forward in [387]. It was based on calculations using Johnson's distribution [369] as the starting point under the following conditions: precipitated water vapor layer 1.0 cm, ozone layer 3.5 mm, 200 aerosol particles of dust per cubic centimeter of air [388]. The resultant flux of the standard terrestrial solar radiation (usually referred to as AM1 radiation) is 917 W/m^2 and the direct component of this radiation is 865 W/m^2.

It is important to note that conditions approaching AM1 are observed in practice only in the tropics and at intermediate latitudes on top of high mountains. Research has therefore continued into the development of the standard spectrum and optimum methods of measurements that would reflect most completely the conditions under which most terrestrial photoelectric installations are used in practice.

A tentative method of testing solar cells for terrestrial applications [389] was developed in 1975 and involves three procedures: natural solar radiation is examined using standard solar cells, non-selective radiometers, and Sun simulators. The instruments and equipment necessary for these tests are described and methods are recommended for calibrating standard cells. The standard provides for illumination under air mass m = 2 and the following atmospheric parameters: thickness of precipitated water vapor 2.0 cm, ozone layer 3.4 mm, turbidity coefficient β = 0.04, and selectivity index corresponding to aerosol absorption α = 1.3 (this spectrum of terrestrial radiation is usually referred to as AM2). The spectral energy distribution of solar radiation under standard conditions was obtained by calculation using the extra-atmospheric spectrum introduced by Thekaekara [356,357]. The standard temperature was taken to be 28 $\pm$ 2°C.

However, the AM2 conditions again do not adequately correspond to the average conditions under which terrestrial cells and batteries are employed, especially in the summer in the southern regions. The tentative method [389] was therefore revised. In the improved procedure [390], the standard conditions were those corresponding to m = 1.5 (AM1.5 conditions). It is assumed that the thickness of precipitated water vapor layer is 2.0 cm, the ozone layer is 3.4 mm, the turbidity coefficient is β = 0.12, and the selectivity index is α = 1.3. The direct flux density in the AM1.5 spectrum is 834.6 W/m^2.

It is proposed that the solar flux density be measured exclusively with standard solar cells.

Of the available artificial sources of light, three are considered acceptable: the short-arc xenon lamp, and the pulsed xenon and tungsten lamps with color temperature at 3400 K and a dichroic interference filter. The absolute radiometric scale must be used in calibration of standard cells.

In addition to the procedures recommended for measurements of the characteristics of solar cells in the direct flux of natural solar radiation, recommendations have also been developed for measurements in the total flux and for cells working with concentrators.

All measurements must be performed in specialized laboratories whose duties include the development of general methodology for the determination of the characteristics of solar cells for terrestrial applications, the calibration of standard cells and their distribution among research organizations, overall metrological control of measurements under manufacturing conditions, and the issuing of the corresponding instructions for measurement.

The metrology of solar cells was subjected to a detailed examination at the Soviet-American Seminar in September 1977, held at Ashkhabad (direct solar energy conversion program; co-chairmen of sessions from the Soviet and American sides were M. M. Koltun and H. Brandhorst).

After a detailed discussion in 1982 by specialists from different countries, including Great Britain, USSR, USA, and France, the procedure for testing solar cells under AM1.5 conditions was adopted as standard by the UN International Electrotechnical Commission [391]. This spectrum is given in Appendix 1 together with the AM0 spectrum [358].

It is important to note that, prior to this decision, the AM1.5 spectrum had been widely used by Soviet specialists in measurements of not only the parameters of solar cells but also of the integrated optical characteristics of materials for solar technology (see, for example, [146]). However, the atmospheric parameters governing the shape of absorption bands were not determined in [146], and these data cannot therefore be used for the purposes of international standardization.

In accordance with the procedure used to measure the parameters of solar cells under natural solar illumination [392], the AM1 conditions were chosen as standard for the total solar flux. In contrast to other studies, where the spectrum of terrestrial radiation was calculated from the extra-atmospheric spectrum, in this procedure the standard distribution of energy in the total flux was determined by averaging eleven experimental curves obtained during four nights in July 1976 on Malta. The latitude of this island is 36° and, at midday in July, the air mass does not exceed 1.03. The standard flux density was taken to be 1000 W/m². Measurements can be carried out in natural solar radiation or in simulated radiation. In the former case, it is recommended that measurements be performed in the total flux with the solar cells and batteries pointing toward the Sun to within ±5°. The flux density should then be not less than 800 W/m². Radiation reflected by the Earth (which can be quite substantial in the presence of snow cover) and by surrounding objects must be excluded. Suitably calibrated standard solar cells are used to measure the natural and simulated solar flux density.

West European countries have recently adopted a common method of testing solar cells for terrestrial applications [366]. This was developed by an international group working under the aegis of the European Common Market at the Joint Scientific Center (IRC) in Italy. Metrological laboratories for the calibration and testing of standard cells for space applications have been established at RAE (Farnborough, UK), CNES (Toulouse, France), and ESTEC (Noordwijk, The Netherlands).

A unified procedure for measurements on solar cells for terrestrial applications [393] has been developed for countries belonging to the socialist economic community. Specialists from Bulgaria, Hungary, Mongolia, Poland, USSR, and Czechoslovakia have been particularly active in the development of this procedure. It includes measurements on solar cells under natural, simulated, and concentrated solar radiation. Steps have been taken to ensure that the procedure is acceptable in the wider, international context, and experience gained in other countries has been taken into account. Two variants of illumination were adopted as standard: $m = 1$, $E_d = 1000$ W/m^2 and $m = 1.5$, $E_d = 850$ W/m^2. The atmospheric parameters are the same in both cases: precipitated water vapor layer 2.0 cm, ozone layer 3.4 mm, humidity coefficient $\beta = 0.12$, and selectivity index $\alpha = 1.3$.

According to this procedure, the parameters of solar cells can be determined in direct and total fluxes.

When measurements are performed in the direct flux, the tested and standard cells must point toward the Sun to within 2° and their field of view must be restricted to 10°. Measurements can be performed under radiation flux densities (measured by a standard cell) of not less than 750 W/m^2 for $m \leqslant 3$.

When measurements are made on the total flux, the tested and reference cells are pointed toward the Sun to within ±5° and are mounted at an angle of not more than 60° to the horizontal plane. The flux density must be not less than 800 W/m^2 and the air mass must not be more than 2. Atmospheric stability, cloud cover, and the albedo of the underlying surfaces are monitored during the measurements in terms of scattered radiation received by the cells: the ratio of the current generated by the reference cell during measurements in the total flux to the current measured in the direct flux must not exceed 1.3. The field of view of the standard cell must be reduced to 10° when the direct flux is measured.

The above procedure was recommended in October 1980 at a conference on the metrology of solar cells, held at Erevan and attended by specialists from countries belonging to the Socialist Economic Community.

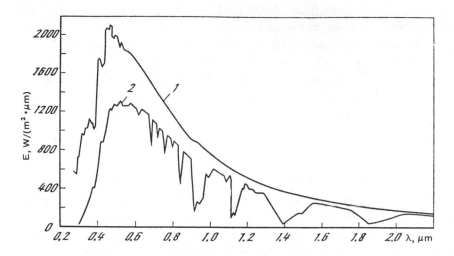

FIG. 4.5. Spectral distribution of extra-atmospheric (1) and terrestrial (2) solar radiation. The latter corresponds to m = 1.5, precipitated water vapor layer thickness 2 cm, ozone layer thickness 3.4 mm, aerosol scattering coefficient α = 1.3, and β = 0.12.

Figure 4.5 shows the currently accepted (in calculations and measurements of the efficiency and output parameter of solar cells and batteries) spectra of extra-atmospheric [358] and terrestrial solar radiation under AM1.5 conditions [390,391].

The importance of standardization of the solar spectrum and of the composition of the atmosphere during measurements can be illustrated by the following example. For AM1.5 and cloudless sky, the direct solar flux density can vary from 943 to 616 W/m^2 [376], depending on humidity and the amount of aerosol particles.

4.2 Measurements on Sun simulators and under natural conditions

The parameters of solar cells and batteries for both space and terrestrial applications are usually measured on Sun simulators and, less frequently, in natural solar light.

Sun simulators are used in various branches of science and technology, for example, in simulating the thermal regime in spacecraft [368], in tests on materials for space applications, medical and biological research, agriculture, sensitometry, calorimetry, and

solar technology. A great variety of optical systems and designs for
Sun simulators has been put forward. Various Sun simulators for
testing space technologies, including solar cells and batteries, are
now available [394].

In the ideal case, the simulator must reproduce with the best
possible approximation all the parameters of solar radiation, i.e.,
the rays must be parallel, the source must be stable in time, and
the illumination, spectral composition, and current density must
be uniform. However, such systems are exceedingly complicated
and expensive and require qualified staff, so that specialized simulators are usually developed for particular applications. In systems
designed for measurements of the characteristics of solar cells and
batteries, less attention is devoted to the collimation of the beam,
and most of the effort is devoted to producing the closest approximation to the solar spectrum and to the stability and uniformity
of the flux. Here again, different approaches can be adopted. When
mass-produced solar cells are examined, the simulator need not
produce the exact solar spectrum, especially when relative measurements are made, for example, in running quality control and sorting
of cells and groups of cells in accordance with their electrical parameters with a view to ensuring low commutation losses after assembly
into batteries. Here, the simulator can have the optimum relationship between complexity of design and precision of measurement.

*High-quality simulators of extra-atmospheric solar radiation
for laboratory studies and sampling measurements*

The sensitivity of most of the widely used solar cells (excluding,
apparently, only cascade systems) is confined to a narrower spectral
range (for example, 0.4-1.4 μm in the case of silicon cells) than the
wavelength range covered by extra-atmospheric solar radiation (0.2-
2.5 μm). This helps in the development of Sun simulators. However,
practically none of the simulators used to test materials for space
applications [368] can be used to measure the parameters of solar
cells because of considerable time and spectral instability of the
simulating radiation. A typical example is the emission spectrum of
the carbon arc, which is quite close to the extra-atmospheric solar
spectrum, but is not stable in time.

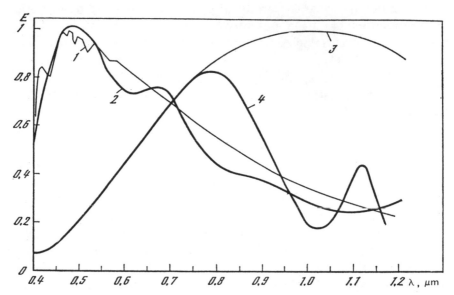

FIG. 4.6. Comparison of the spectral energy distributions produced by simulators with extra-atmospheric solar radiation: 1) extra-atmospheric solar radiation; 2) S-1 simulator; 3, 4) hot-filament lamps before and after correction by a 4-cm water filter, respectively.

Sun simulators capable of reproducing the solar spectrum to a high degree of precision, and generating a uniform flux, are used in research and in sampling measurements on manufactured and prototype solar cells. Uniform illumination is produced by mixing beams of rays by one of a number of available methods. The S-1 simulator manufactured in the Soviet Union incorporates a hot-filament lamp with a color temperature of 3100 K [395,396]. It produces a flux of radiation that is uniform to within ±10% (over an area of 20 X 30 mm) by superimposing two beams of radiation. Spectral correction is introduced by colored optical glasses (SZS-14, 1 mm thick, SZS-17, 3.5 mm and PS-14, 8 mm). Light filters are used to reproduce the spectrum with good precision in the range 0.4-1.1 μm (Fig. 4.6), but the filters themselves absorb a substantial fraction of the radiation produced by the lamp, so that its output must exceed by a factor of ten the simulated flux, and the light filters must be efficiently cooled. The attendant difficulties have been overcome [395] by using a 750-W hot-filament lamp with a two-beam system and a

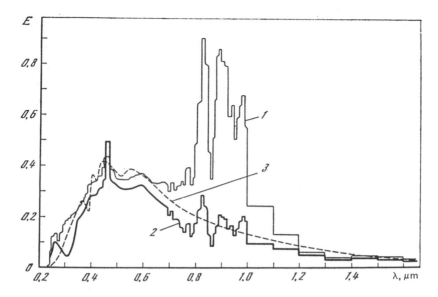

FIG. 4.7. Emission spectrum of a high-pressure xenon lamp: 1, 2) before and after correction with interference light filters, respectively; 3) smoothed extra-atmospheric solar spectrum.

special cooling arrangement (the light filters are immersed in transparent carbon tetrachloride cooled by flowing water).

The spectrum produced by the S-1 simulator is monitored by measuring the blue/red ratio, i.e., the ratio of short-circuit currents produced by a standard solar cell when filters transmitting blue/green and infrared radiation (SZS-18 and IKS-21 filters, respectively) are successively introduced into the beam. The radiation flux density can be varied without affecting its spectrum by using variable-aperture diaphragms and neutral or wire grid filters.

Highly uniform illumination is usually produced in accurate simulators by using a special mixer — the optical integrator [397] — in the form of a stack of hexagonal lenses. An image of the light source is projected onto the front face of the stack. The integrator consists of a large number (up to 19) of individual projection systems, each of which produces its own beam and directs it onto the entire working area on which beams from all the elements in the mixer are superimposed. In contrast to the usual projection system, in which the nonuniformity of the light source is reproduced on

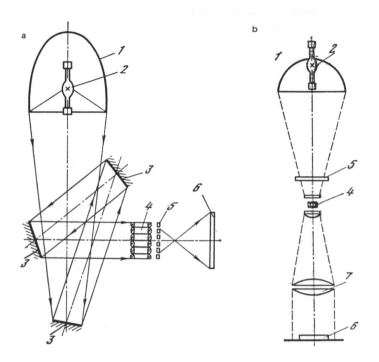

FIG. 4.8. Optical system of the Spectrosun X-25 simulator (a) and the small Ushio Electric simulator (b): 1) large-angle reflector; 2) high-pressure xenon lamp; 3) flat mirror; 4) integrator-mixer; 5) interference filter; 6) solar cells under test (or groups of them); 7) objective.

the illuminated surface, the primary image is fragmented and a large number of light spots due to all the elements in the mixer is superimposed. The intensity on the illuminated surface is found to be uniform to within ±2-3% of the average.

In most modern simulators, the rays are made parallel by collimators (usually paraboloidal mirrors or Fresnel lenses). An image of the light source is formed in the focus of these collimators by suitable condensers (usually, large-aperture ellipsoidal mirrors). The decollimation angle is equal to the ratio of half the ray-beam diameter at the collimator focus to its focal length [394].

The source of radiation in most simulators manufactured outside the USSR is a high-pressure xenon lamp. The spectrum is cor-

rected by interference filters, so that the spectrum of the lamp is
brought as close as possible to the extra-atmospheric solar spectrum
[358] (Fig. 4.7).

The Spectrosun X-25 simulator manufactured by Spectrolab
(USA) was developed for measurements on solar cells. It produces
a beam that is uniform to within ±2% over an area 300 mm in
diameter at a distance of 1.5-2 m from the cassette containing the
light filters [21,398,399]. The demountable set of filters can be
used to produce both extra-atmospheric and terrestrial solar spectra,
although the latter is very different from the standard spectrum
(AM1.5 conditions).

Similar principles are exploited in simulators manufactured by
Ushio Electric (Japan), Oriel (USA), Optical Radiation Corporation
(USA), Bosh (FRG), and so on. Figure 4.8 illustrates the optics of
the Spectrosun X-25 and the small simulator manufactured by
Ushio Electric.

Among the Soviet simulators with intermediate working area,
the instrument developed at the All-Union Optical Technology
Institute has good parameters [400]. The illumination is uniform
to within ±2% over an area of 150 X 200 mm. It is produced by a
mixer in the form of a relatively long (1-2 m) vertical hollow mirror
lightguide, whose cross section is just greater than the working area.
However, the simulator does not produce the highly parallel rays
typical of extra-atmospheric solar radiation. It incorporates two
metal-halogen lamps with a near-solar spectrum and mercury gas-
discharge lamps containing traces of tin iodide and bromide [401,
402]. Emission spectra of metal-halogen lamps filled with tin bromide, aluminum chloride, and indium iodide are reported in [368].

It is important to note that accurate simulators of extra-atmospheric solar radiation incorporating components whose optical
characteristics vary rapidly in time and have to be changed regularly
(e.g., multilayer interference light filters, complicated lamps with
envelopes whose transmission deteriorates in time, and whose radiation parameters are not constant) are unsuitable for quality control
in the manufacture of solar cells. Moreover, such simulators are not
designed for measurements on solar batteries which usually have a
large area (some tens or even hundreds of square meters).

Simulators for measurements of the parameters
of mass-produced solar cells and batteries

The simplest simulator that is convenient under industrial conditions and is stable consists of hot-filament tungsten lamps with mirror or diffuse reflectors arranged so as to illuminate solar-cell panels of practically any area [403].

A large proportion of the infrared radiation produced by such lamps (which heats the solar cells during measurements) can be removed by heat-reflecting filters, placed between the lamps and the cells. They consist of glass plates coated with a transparent conducting film of tin dioxide, a mixture of tin dioxide and indium trioxide, or cadmium stannates with surface resistivity of less than 50 $\Omega/\square$ (the film must be deposited on the glass plate facing the lamp).

The infrared component can be reduced still further by using a 2-4 cm layer of water as the heat-absorbing filter. The water filter itself can be cooled by using an external radiator, or by flowing water, and air bubbles can be removed from overheated water by mechanical brushes [404]. The effect of the water filter on the spectrum of the hot-filament lamp can be seen by comparing curves 3 and 4 in Fig. 4.6.

Such simple simulators with a water filter are widely used in rapid quality control of solar cells and groups of cells (up to 20 X X 30 cm) at all stages of the fabrication process or, without the water filter, in quality control of solar batteries.

The spectrum of hot-filament lamps used for quality control in the manufacture of large-area solar batteries can be corrected and brought closer to the solar spectrum by depositing multilayer interference light filters on the inner surface of the lamp bulb (both in front of and behind the hot tungsten filament) [405,406]. The bulb protects the filters from the unfavorable influence of the ambient medium (e.g., enhanced humidity), while the thermal effects due to the tungsten radiation, which give rise to the crystallization of the filter material and its subsequent detachment from glass, can be avoided by inserting a thin semitransparent chromium film between the layers in the filter. This film must be deposited rapidly in a high vacuum [406]. The deposition of a layer of tungsten on the interior of the glass bulb and on the filters can also be prevented by using a headlight-type lamp with the light filter deposited on its glass bulb

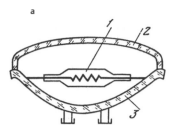
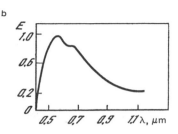

FIG. 4.9. Headlight-type lamp (with multilayer interference light filters deposited on the inner wall) for Sun simulators (a) and its emission spectrum (b): 1) KGM-110-500-2 built-in quartz halogen lamp (500 W, 110 V); 2) reflector coated with a thin-film multilayer light filter; 3) scatterer with a multilayer color-correcting filter deposited on heat-reflecting tin dioxide film.

and a powerful tungsten lamp in a quartz bulb placed in its interior [407]. Figure 4.9 illustrates the design of a tungsten headlight for Sun simulators with interference filters on the interior of the walls. The spectrum of the radiation produced by this lamp is also shown. These headlight-type lamps can be used in a Sun simulator in measurements of the parameters of solar batteries of any area.

A 300-W hot-filament lamp of similar design with an internal interference filter (operating voltage 120 V) has been developed by the General Electric Co. (model ELH). These lamps, working in conjunction with Fresnel lenses mounted in front of them and producing a parallel beam, have been used to develop a convenient and simple Sun simulator producing a uniform flux over an area of 1.2 × 1.2 m at a distance of 4.6 m [408]. The simulator incorporates 143 quartz halogen lamps with an elliptic reflector. A multilayer interference coating (dichroic mirror) is deposited on its surface and transmits a large proportion of the infrared radiation and reflects visible radiation. There are also 143 hexagonal Fresnel lenses, mounted at 28 cm from the lamps. By varying the supply voltage for the simulator lamps, it is possible to vary the density and spectrum of the emitted radiation between relatively wide limits.

Pulsed xenon lamps [409-411] are usually used to illuminate large areas (3 × 3 m or more). These simulators have no optics, and uniform illumination is achieved by placing the battery under examination at a large distance from the lamp. The spectrum is corrected by an interference or, occasionally, a water filter. It is very important to provide the simulator with the necessary measuring equipment which must ensure that all the points on the current-voltage characteristic of the battery can be recorded during a single pulse of about 1 msec. Simulators of this type include the LAPSS simulator for measurements on space batteries. It produces uniform illumination to within ±2% on an area of 2.5 X 2.5 m [409]. A similar principle is used in the small TTPSS vertical-type simulator (2.3 m high) which also produces uniform illumination to within ±2% over an area of 0.6 X 0.6 m. The simulator is relatively cheap, but only one point on the current-voltage characteristic can be obtained per pulse [412].

When a solar battery is examined under a pulsed simulator, it is not heated, and its temperature remains close to the room value [21].

The standard temperatures used in different countries for calibration purposes are 40, 28, and 25°C. In the USA and Western Europe, 28°C has been adopted as standard [389-393]. This is hardly a successful choice, since cells and batteries are usually heated by solar radiation, and the temperature of 40°C corresponds more closely to the operating extra-atmospheric and terrestrial conditions.

When measurements are made under pulsed simulators, a computing device automatically converts the battery parameters to the working temperature. This conversion is based on the average temperature coefficients which are subject to considerable spread. It is probably better to provide pulsed simulators with a thermostat controlling the temperature of the solar cells under examination, and monitor the temperature of the battery at the time of measurement. A temperature-stabilizing device can be based, for example, on infrared sources which are mounted on the dark side of the battery while the measurements are carried out.

In conclusion, we must briefly consider continuously operating ultrapowerful xenon lamps, each of which can produce the necessary radiation flux of 1360 W/m^2 on the surface of the solar battery with an area of a few tens of square meters (and produce an acceptable simulation of the extra-atmospheric solar spectrum). A source of this kind has been developed at the All-Union Light Technology Institute. It takes the form of a high-pressure metal xenon lamp generating 40 kW [413]. The lamp is explosion-proof, and is provided with a water-cooled double quartz window in a metal body.

However, the illumination is nonuniform, falling by ±20% at the edges of the working area [368], so that such lamps can only be used in systems designed to investigate light aging in space, or in crude tests on the state of solar batteries. They are not suitable for measurements of photoelectric parameters.

Simulators of terrestrial solar radiation

The complex character of terrestrial solar radiation for different values of the air mass (see Fig. 4.2) means that it is very difficult to simulate this radiation even when we confine our attention to reproducing the standard AM1.5, AM2, and AM1 conditions [387, 389,391] in the wavelength range 0.4-1.1 μm. The standard terrestrial solar radiation can probably be reproduced only by using a monochromator with a program-controlled entrance slit, but this means that intensities typical for solar radiation cannot then be produced even with a high-luminosity monochromator. Another possible approach is to use a xenon or halogen lamp equipped with a set of demountable narrow-band interference filters to define narrow spectral ranges. Either method will produce the simulated solar flux over a small area (a few square millimeters or centimeters).

Because the precise simulation of terrestrial solar radiation is so difficult to achieve, approximate methods are widely used, and simulators producing the smoothed average radiation curve under AM1.5, AM2, or AM1 conditions have been developed. They incorporate suitably chosen light filters. Many of these simulators provide a good enough approximation to terrestrial solar spectra when the filters and the direct flux density are suitably chosen. For AM2, AM1.5, and AM1, the flux density should be close to 750, 850, and 910-950 W/m^2, respectively.

As an example, Fig. 4.10 shows the radiation curve for a simulator incorporating halogen lamps [408], together with the AM2 standard solar spectrum. To produce a direct flux density close to 750 W/m^2, the working voltage across the halogen lamp with the built-in dichroic filter (type ELH) [408] was reduced to 100 V.

A simulator for measurements of solar-cell parameters is described in [414] and consists of two lamps — a xenon and a tungsten lamp. The long-wave part of the emission of the xenon lamp (to the right of 0.7 μm) is "cut off" by a filter based on a copper sulfate solu-

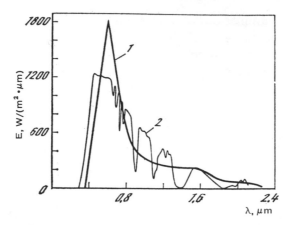

FIG. 4.10. Simulation of the standard solar spectrum under AM2 conditions: 1) simulator consisting of halogen lamps with built-in dichroic interference filters and Fresnel lenses; 2) standard spectrum of terrestrial radiation under AM2 conditions.

tion cooled with water, while the short-wave radiation emitted by the hot-filament tungsten lamp (to the left of 0.55-0.6 μm) is absorbed by a colored glass filter. By superimposing two beams of radiation corrected in this way on an area of 1 X 2 cm, and by varying the intensity of the lamps and the thickness of the filters, it is possible to produce a smoothed curve similar to that of extra-atmospheric or terrestrial radiation.

Useful practical results have been obtained by comparing the spectra of two simulators of terrestrial solar radiation incorporating xenon and tungsten lamps (with partial filtration of the radiation) with standard AM1 solar radiation [415]. Comparison of the parameters of thin-film copper sulfide—cadmium sulfide solar cells, measured under natural conditions close to AM1, with parameters obtained in the laboratory under both types of simulator, has shown that the xenon-lamp simulator provides a much better approximation to the working conditions of solar cells. This is probably due to the absence of the short-wave radiation ($\lambda < 0.4$ μm) in the case of the tungsten lamp simulator, since the sensitivity of solar cells in this region is still quite high.

Simulators incorporating continuously operating and pulsed xenon lamps are much more expensive because they have to be

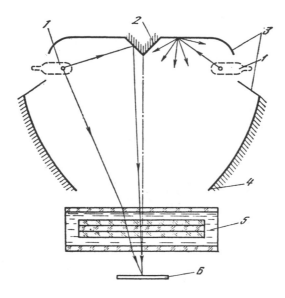

FIG. 4.11. Optical system of terrestrial simulator with diffuse reflector: 1) halogen tungsten lamp (KGM-30-300-2); 2-4) faceted mirror, diffuse, and spherical mirror reflectors, respectively; 5) composite water-cooled filter consisting of a set of colored glasses; 6) solar cell under examination.

equipped with demountable sets of multilayer interference filters to reproduce terrestrial solar radiation. Simpler terrestrial simulators for measurements of the parameters of silicon and gallium arsenide solar cells can be based on hot-filament lamps with more stable parameters, and do not require complicated stabilization of supplies. These solar cells have low sensitivity for $\lambda < 0.5$ μm (the sensitivity is practically zero in this region when the depth of the p—n junction in silicon is large and the thickness of the wide-gap AlGaAs filter on the surface of gallium arsenide is large). The parameters of these solar cells can be measured with sufficient precision under simulators based on hot-filament tungsten lamps with colored-glass filters. One of these simulators [416] incorporates a 650-W halogen lamp working at a color temperature of 3200 K. The spectrum is corrected with the CSI-75 Corning Glass filter, and a flux density of 740 W/m^2 can be produced in a spot 6 cm in diameter. The illumination is uniform to within ±1%.

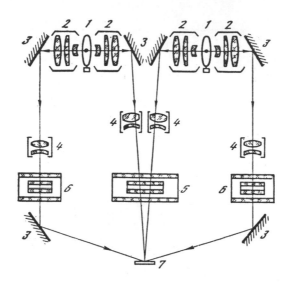

FIG. 4.12. Optical system of a simulator of direct and scattered (diffuse) terrestrial solar flux: 1) halogen tungsten lamps; 2) condensers; 3) faceted flat reflectors; 4) objectives; 5, 6) filters for the simulation of direct and scattered (diffuse) flux of radiation, respectively; 7) solar cell.

A relatively simple terrestrial simulator incorporating hot-filament lamps, glass filters, and a diffuse reflector is described in [366]. It produces uniform illumination by scattered radiation, similar to that observed under field conditions (Fig. 4.11). Experiments have shown that this reflector produces uniform illumination to within ±5% on an area of 40 × 40 mm. The simulator does not contain any lens optics. The radiation is emitted by halogen lamps with a color temperature of 3400 K. The spectral distribution of the total flux of terrestrial radiation under an air mass of 1.5 can be produced by using the SZS-24, SZS-17, and PS-14 colored glasses.

A better approximation to the terrestrial radiation can be achieved with the optical system shown in Fig. 4.12 [366]. The right-hand beam from one lamp and the left-hand beam from the other pass through a light filter and, by illuminating the solar cells at an angle close to normal incidence, simulate the direct flux of solar radiation. The other pair of beams passes through the correction system and enters the solar cells at an acute angle, simulating the scattered sky radiation. Calculations have shown that the spectral distribution of the hot-filament lamp with a color temperature of 3400 K can be transformed into the spectral distribution of direct solar radiation with standard parameters by using a filter consisting of the following colored glasses: SZS-17 (2 mm), SZS-15 (1 mm), PS-14 (3.2 mm), and a distilled water layer of 25 mm. The filter thickness was optimized on a computer, and good correction of the lamp spectrum was achieved.

*Measurements of the parameters of solar cells
and batteries in natural solar light*

The development of solar energy conversion systems must be preceded by field tests on terrestrial solar batteries under practical conditions. Natural terrestrial radiation can also be used in grading tests on both terrestrial and space batteries. The radiation flux density is then, of course, determined with a standard solar cell calibrated for the corresponding standard conditions. It is important to emphasize that, when standard solar cells are employed, terrestrial solar radiation will be the same under certain specific conditions as the standard radiation (terrestrial and space). For example, when a collimator is used, the terrestrial solar radiation at midday in the summer, at a high altitude, dry atmosphere, and small amounts of aerosols, is closer to the radiation in space than anything produced by even the best simulators. The spectral energy distribution of terrestrial solar radiation under an air mass m = 1.03 has been measured on a Californian plateau at an altitude of about 2300 m above sea level (Fig. 4.13) [417]. Over a period of 1 min, the flux density of the radiation at this altitude was found to vary by not more than 0.5%. Very similar conditions obtain at the high-altitude station of the Shternberg Astronomical Institute (about 3000 m above sea level), where standard solar cells are calibrated annually [418].

The standard spectral energy distribution of terrestrial radiation has been calculated using the average concentration of all the components of the atmosphere. These conditions are not typical for any of the seasons of the year, but natural radiation is frequently closer to the standard than the radiation produced by high-grade terrestrial simulators.

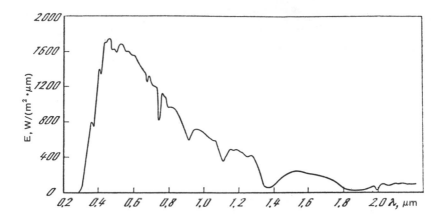

FIG. 4.13. Typical spectral distribution of terrestrial solar flux density under high-altitude conditions (2300 m above sea level, atmospheric pressure 585 mm Hg, thickness of the layer of pre-cipitated water vapor 5 mm, ozone layer 2.5 mm, dust particle concentration in air 200 cm^{-3} ; m = 1.03).

The parameters of terrestrial cells and batteries can be mea-sured in three ways in natural radiation, namely, by placing the bat-tery horizontally, mounting it at an angle to the southerly direction, approximately equal to the latitude of the particular locality, or using a tracking system so that the normal lies along the direct flux. The standard solar cell must always be mounted in the same plane as the battery under examination.

Each method has its advantages and disadvantages. The first of the three is the simplest and prevents the incidence of radiation reflected from the ground, but measurements can only be carried out when the Sun is high above the horizon [392]. The second method is closer to the conditions under which solar batteries are actually used in practice, but requires shielding from radiation reflected by the underlying surface. The third method is usually employed with collimating tubes. It gives the most reproducible results, but requires the use of a tracking system and collimating devices, so that it cannot be used with large batteries.

Whichever method is used, the radiation flux density must not differ from the standard value by more than 20%. This keeps the uncertainty in the calculated parameters of cells and batteries under

4. Determination of the efficiency and
metrological parameters of solar cells and batteries
227

standard conditions at an acceptable level. The temperature of the cells and batteries during such measurements must be kept near the standard value. This is readily achieved for individual cells, but uniform thermal stabilization of batteries is often difficult. Accurate measurements of the working temperature are frequently used instead. The battery must then be well protected from wind, so that the temperature is uniform over the entire area. The cell and battery parameters at standard temperature and under standard illumination are calculated from the formulas given in [392].

4.3 Standard solar cells

Since the spectral energy distribution of even high-grade simulators is not identical with the standard solar distribution, and the sensitivity of solar cells is selective, the simulator intensity cannot be adjusted with the aid of nonselective radiation detectors (radiometers). Specially calibrated standard solar cells [419] must be used for this purpose. These standard cells, often referred to as photometer devices, are actually radiometers with selective sensitivity.

Calculations have shown that, for a given air mass, relatively small variations in the principal components of the atmosphere can be accompanied by quite substantial variations in the solar flux density [376]. Comparison of different atmospheric conditions has shown that while the solar flux density measured by a *nonselective* radiometer may be the same on a number of occasions, the corresponding spectral compositions may be so different that solar cells (by virtue of their *selective* sensitivity) will deliver different electric power and very different photocurrents. Even in the case of high-grade cells, the difference between short-circuit currents measured under terrestrial conditions with the same spectral energy density but different states of the atmosphere may amount to 15% [392]. The same solar radiation density of 672 W/m^2 can be observed for the following two states of the atmosphere: m = 1.5, ozone layer thickness 2 mm, β = 0.17, α = 0.66, and m = 3, ozone layer thickness 5.5 mm, β = 0.02, α = 1.3 (thickness of the layer of precipitated water vapor is 2.0 cm in both cases). It is clear, however, that the spectral composition of the radiation can be appreciably different for such different atmospheric parameters.

Comparison of the calibration factor, i.e., the ratio of the integrated photocurrent per unit area of the cell (determined from the spectral sensitivity) to the solar flux density incident on the same area, for a large number of solar cells has shown that, when the intensity produced by a simulator consisting of tungsten lamps without filters is adjusted against a nonselective radiometer, the uncertainty in the measured short-circuit current of solar cells may be as high as 50% [420]. When simulators incorporating tungsten lamps and a dichroic filter are employed, the uncertainty is 30% (when predicting the current in space) and 10% (under terrestrial conditions), while for simulators incorporating short-arc xenon lamps and an interference filter, the uncertainty is 15% for terrestrial measurements and 3-5% for space measurements.

Calibrations of standard solar cells involve measurements of the short-circuit current under standard illumination. A standard solar cell is used to adjust the simulator, i.e., its radiation flux is adjusted until the short-circuit current of the standard cell is the same as under standard conditions.

It is important to note that the incident solar flux density on the working area of the simulator is not then exactly the same as the corresponding quantity produced by natural solar radiation under standard conditions because the radiation is estimated from its effect on a selectively sensitive solar cell of a particular design and of a particular semiconducting material.

It is common to introduce special units when the level of radiation is estimated from its effect on a receiver with a particular spectral sensitivity: its effect on the human eye is expressed in terms of lux, its effect on the skin in terms of erythemic units, and so on. However, this involves equivalent rather than effective quantities. Thus, when a source with arbitrary spectrum producing a given incident energy flux density generates a current in a solar cell equal to that produced by extra-atmospheric radiation, the equivalent flux density for the given cell is 1360 W/m². For example, when a solar cell with a shallow p–n junction ($l \leqslant 0.5$ μm) is illuminated by a hot-filament lamp with a color temperature of 2850 K, the current it generates is the same as that generated in space if the incident energy flux density produced by the lamp equipped with a 40-mm water filter is 780 W/m², while the corresponding figure for the lamp without the filter is 960 W/m². In both cases, a standard silicon cell will produce a reading indicating 1360 W/m^2.

When a standard solar cell is employed, satisfactory precision of measurement can be achieved even with poorly corrected simulators or sources with arbitrary spectral energy distribution. The uncertainty in the measured electrical parameters of a tested solar cell will then depend on the extent to which its spectral sensitivity differs from that of the standard cell. It is thus clear that the basic requirement imposed on standard solar cells is that their optical and, especially, spectral characteristics must be identical with the characteristics of the solar cells with which they are compared. When standard cells are employed under simulators producing a wide beam of radiation, it is also necessary to take into account the angular distribution of sensitivity. This is largely determined by the surface microtopology of the solar cell, which influences the reflection coefficient as a function of angle of incidence [23]. Even the most advanced technological process will not result in identical optical and spectral characteristics among all the cells of a given type. Standard cells must therefore be chosen so that their characteristics are as close as possible to the average characteristic for a particular series.

The design of standard solar cells includes the development of the physical configuration, investigation of stability and metrological parameters, and the development of calibration equipment and methodology.

Structure of standard solar cells

The structure of standard solar cells depends on their application but, in all cases, it must satisfy the basic requirement that all the cell parameters must be very stable (as for all measuring devices). This, in turn, necessitates reliable thermostating of solar cells or the precise measurement of their temperature. The simplest is the solar cell mounted in a recess in a metal plate and protected by glass [390,419,421]. Constant temperature is ensured by placing the standard cell on a thermostated table.

The new standard cells PS-4, -5, -6, and -7 [422] are silicon solar cells. The PS-4 has a photosensitive element of 14 × 24 mm with a contact strip 2 mm wide all over its periphery. The element is held against a silvered brass body by comb-type springs. They are also used as current contacts for the upper electrode and have a low contact resistance (not more than 0.03 Ω). The optical glass cover isolates the solar cell from the ambient medium. The temperature is held at the required level by a heating element built into the body of the cell and two temperature sensors (actuating and monitoring) mounted directly on the back of the cell. The whole system operates automatically, and the temperature is held constant to within ±1°C.

In the PS-5 cell, the silicon element (10 × 20 mm) is soldered directly to the body and is protected by glass. Both the glass and the electrical leads are suitably sealed to the body. The system is assembled in a dry inert gas atmosphere to ensure stability of its working characteristics. Constant temperature is maintained by mounting the standard cell on a thermostated table. Water from a liquid-filled thermostat circulates in the interior of the table. A calibration curve is provided, giving the open-circuit voltage as a function of cell temperature for a given incident flux density.

The PS-6 incorporates a silicon solar cell of 10 × 20 mm, side by side with a silicon temperature sensor whose optical and thermophysical properties are identical with those of the cell except that it does not contain the p—n junction. Both are protected by glass.

The PS-7 consists of a group of solar cells with an overall sensitive area of 76 × 71 mm. The temperature sensor is located at the back of the group. The cells are protected by a common glass shield fixed to the body of the receiver (this facilitates servicing).

The PS-9 was developed in the USSR in 1980-1982 as a standard for countries in the Socialist Economic Community. It has a rectangular photosensitive area of 30 × 35 mm or a circular area with a diameter of 50 mm for measurements of the parameters of cells and batteries for space [423] and terrestrial [424] applications, respectively.

This new standard cell has a built-in cooling system incorporating a radiator, which can be cooled with water from a thermostat, and a sensitive temperature sensor. The PS-9 uses silicon solar cells with a shallow p—n junction and heterojunction cells consisting of a solid solution of aluminum in gallium arsenide and gallium arsenide. The large body of the PS-9 ensures a field of view in excess of 166°, which means that it can be used in measurements on solar cells and batteries in both total flux and direct collimated flux.

Provision is made in the new standard cell for a tube that

reduces the field of view to ±2.5%, which is necessary when the direct solar flux is measured in determinations of the characteristics of batteries working in conjunction with concentrators. The parameters of the atmosphere (concentration of water vapor, ozone, and aerosols) can be monitored by inserting interference filters into the tube. They transmit radiation in narrow spectral intervals corresponding to selective absorption bands in the spectrum of terrestrial solar radiation.

The absolute calibration of standard cells is laborious, time-consuming, and costly. Standard cells that have been thus calibrated are therefore used only in primary measurements. Working photometers are used for every day purposes.

The PS-4 and PS-9 standard cells were developed as primary standards. The following are measured when these cells are evaluated: the current-voltage characteristic, the temperature dependence of the short-circuit current in the range between 20° and 60°C, the short-circuit current as a function of the incident energy flux density for solar constants between 0.5 and 1.5, and the variation in stability over a period of not less than one year. These measurements are then used to select a group of cells whose sensitivity is stable to within 0.5%, and whose temperature coefficient of the short-circuit current is not more than 0.2%/°C.

The PS-6 and PS-7, whose sensitivity does not change by more than 1.5%, are used as working photometers. They are calibrated against standard cells on simulators similar to those for which they are designed. All the working cells are checked against standard cells at least once every six months.

The angular field of view of a standard cell has no effect on the precision of measurement when it is used in a collimated beam. However, it is often important, for example, in measurements of the total flux density under terrestrial conditions, to ensure that the cell body does not cast a shadow on the sensitive area. The 166° standard cell can be used for all possible values of the scattered solar flux and any type of cell surface treatment [390].

Constant improvements in fabrication technology and the development of new types of solar cell mean that the parameters of solar cells with nonstandard spectral sensitivity have to be measured. This demands the availability of standard solar cells with different spectral characteristics. Solar cells for these standards

are produced by altering the depth of the p—n junction, by varying
the properties of the antireflective coatings, and by bombarding the
cells with protons and electrons of different dose and energy. A
quick test can be based on the comparison of the blue-red ratios of
the standard cell and the cell to be examined, using filters to isolate
the blue and the near-infrared radiation, respectively. The standard
cell with the nearest value of the blue-red ratio is then selected.
This approach can also be used to select standard cells when the
parameters of solar batteries consisting of nonstandard solar cells
are measured. A set of standard cells with nonstandard spectral
sensitivity distributions has been developed on the basis of the PS-5.

Stability of the characteristics of standard solar cells

Standard cells are selected from among mass-produced cells,
or are fabricated specially. When they are selected, particular attention is paid to the quality of the end surfaces, the shunt resistance,
and the series resistance. They must have uniform properties (especially the spectral and integral sensitivities) over their working area.
It is desirable that they have the minimum possible temperature
coefficient of short-circuit current. Cells selected in this way are
mounted on frames and subjected to natural or accelerated aging.
Their stability is examined over a long period of time, and the
methodology employed must ensure that relative measurements
are made with a precision of at least 0.1%. Primary calibrations
employ standard cells whose sensitivity remains constant to within
±0.5% [419]. The angular distribution of sensitivity and the linearity
of the light characteristic are also checked for standard solar cells
for terrestrial applications. The deviation from linearity in the range
400-1000 W/m^2 must not exceed ±0.5% of the nominal flux density
under AM1.5 conditions.

Silicon solar cells designed for energy conversion and used as
standards have the most stable characteristics of all solar energy
converters under ordinary conditions. Their short-circuit current
is a linear function of the radiation flux density in a relatively
broad range, and they have a small temperature coefficient of short-
circuit current (0.1-0.2%/°C). Their sensitivity covers the visible and
near-infrared ranges of the spectrum. It is possible to produce silicon
solar cells with an ultra-shallow p—n junction (thickness of the top

doped layer $l \leqslant 0.1$-0.2 μm), which are sensitive to the solar near-ultraviolet.

However, by far not every mass-produced solar cell will have the parameters (and the stability) expected of measuring devices. It has therefore been necessary not only to develop standard solar cells with the required stability and reliability under working conditions, but also to perform careful preliminary selection of cells for standards. Another important aspect was the mounting of solar cells within the standards.

The stability of cell sensitivity is determined by measuring the short-circuit current after the cells have been fully assembled. Since 1967, such measurements have been carried out regularly at the rate of at least twice per annum. A total of 75 standards was prepared in 1967, and work began on the determination of the stability of their sensitivity. Cells that cease to work satisfactorily, or those that were found to have unstable sensitivity, are periodically discarded, and new sets of standards consisting of improved silicon cells, gallium arsenide homojunction cells, and AlGaAs–GaAs and Cu_2S–CdS heterostructures have been added to this set. Standard solar cells whose properties turned out to be stable over a period of one year or more (the short-circuit current remained constant to within $\pm 0.5\%$) were then chosen as primary standards. They were calibrated directly in solar radiation or by comparing them with primary standards of flux density.

The stability of sensitivity was checked against Sun simulators whose spectrum, uniformity, and, especially, flux density were carefully monitored. To estimate the stability of sensitivity to within 0.5%, the simulator flux density must be monitored to better than 0.3%, which presents a serious difficulty. The light source is supplied by an electronically controlled voltage stabilizer, the electric power output of the lamp is monitored by a wattmeter of precision class 0.1, and a continuous-flow thermostat is used to maintain the temperature of the solar cell in the body of the standard to within $\pm 0.5°C$.

Figure 4.14 shows the measured short-circuit current of primary standard cells as a function of time. Standard 1 was made in 1962 and incorporates a solar cell without coatings. It was produced by thermodiffusing phosphorus into single-crystal p-type silicon [122, 125]. Standards 2-6 with antireflective coatings [290] were

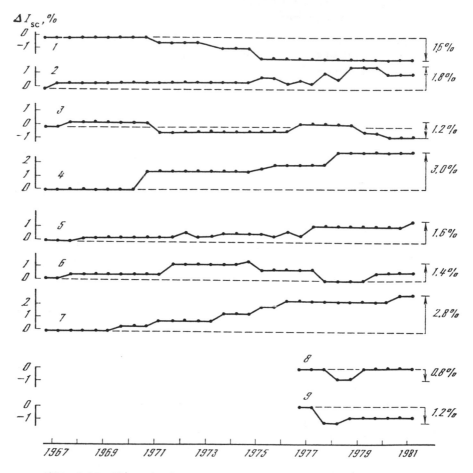

FIG. 4.14. Short-circuit current of standard silicon solar cells measured as a function of time between 1967 and 1981. 1-9) Cell numbers.

made in 1967. Out of the 75 standard cells assembled in 1967, seven retained their sensitivity to within ±1.5% by 1981, despite the fact that none had glass shields. The standard solar cells were kept in a laboratory without any special precautions, and were used during the periodic checks two or three times a year daily for one or two weeks.

It is clear from Fig. 4.14 that the short-circuit current of standard solar cells can either increase or decrease in the course of time. For example, cell 4 was found to increase its sensitivity by 3% over a period of twelve years. Standard cells for which the short-circuit current was found to vary by more than 0.5% over a period of six months were transferred to the working set.

In a substantial proportion of the standard cells, sharp changes in sensitivity were found to be unrelated to external mechanical defects (the cells were examined under a microscope) [425].

In working standards, the receiver of radiation can be either an individual solar cell or a group containing 60-70 cells [419]. The stability of the sensitivity of these standard groups is largely determined by the way in which they are used (daily application with frequent changes of illumination and temperature, possible heavy mechanical wear). The working surface of the solar cells in these standards has an antireflective coating, but is not protected by glass. It is arranged flush with the body. Figure 4.14 shows the short-circuit current of a number of standard groups as a function of time. The short-circuit current of these groups is found to vary more rapidly and frequently than that delivered by single solar cells. However, prolonged use of these cells has shown that more than 90% of them retain their sensitivity to within ±1.5%.

These results show that the semiconductor solar cell (suitably protected from the ambient medium) is one of the most stable devices for measuring optical radiation.

4.4 Calibration of standard solar cells

Standard cells are operated under short-circuit conditions and their calibration involves the determination of the short-circuit current under exposure to normalized solar spectrum and flux density (extra-atmospheric or terrestrial). Two basically different types of calibration are possible, i.e., calibration in natural solar radiation or in the laboratory, using a methodology consistent with the USSR State Standard (GOST 8-196-81). A variety of methods belonging to the first type is used to calibrate standard cells for simulators of extra-atmospheric solar radiation. They include measurements from spacecraft [362], rockets [426], sounding balloons [427], and high-flying aircraft [428], as well as measurements on the ground [414,429,430].

In the case of spacecraft and rockets, the short-circuit current is measured directly under extra-atmospheric conditions. Calibration on spacecraft is not only expensive, but also subject to a number of complications related to the return of the standard cells to the Earth, and are therefore usually employed only to check the validity of other methods. Rockets, ascending to a height of more than 200 km, return the standard cells to the Earth. All measurements are performed at a height of at least 100 km [426].

Sounding balloons are sent to 30-40 km, so that the spectral energy distribution of solar radiation is then determined almost exclusively by the ozone absorption bands and, to a much lesser extent, by aerosol scattering. The influence of ozone and of aerosols is taken into account by introducing a suitable correction.

Aircraft used in scientific experiments usually ascend to 12-13 km. The standard solar cells are made to face the Sun by the pilot, using an optical sighting system. Measurements begin at 3-4 km. The solar radiation parameters depend on the altitude of the aircraft above sea level and on the position of the Sun above the horizon, i.e., on the optical air mass. Aircraft experiments have been carried out under absolute air masses of between 1.4 and 0.14 [362]. The current corresponding to extra-atmospheric conditions was determined by extrapolating the results to zero air mass. The same procedure can be used with direct measurements performed under terrestrial and, preferably, high-altitude conditions.

High-altitude measurements

The method most frequently used in calibration measurements under natural solar illumination on the ground (usually at high altitudes) involves extrapolation to zero air mass [431]. Calibration involves successive measurement of the short-circuit current delivered by standard solar cells for different values of the air mass (different positions of the Sun above the horizon). Since the experiments are performed under stationary conditions, it is sufficient to determine the short-circuit current as a function of the relative air mass. The extra-atmospheric value of the short-circuit current is obtained by linear extrapolation of the logarithm of the current as a function of relative air mass to its zero value.

In practice, the short-circuit current is measured for half the solar day. The logarithm of the measured current is plotted as a function of the air mass, and a straight line is drawn through the experimental points (the so-called Bouguer line), which is then linearly extrapolated to the current at zero air mass. Strictly speaking, the relationship is linear only for monochromatic light. Since the silicon solar cell is sensitive in a sufficiently wide spectral region, the Forbes effect ensures that this plot is a slightly concave curve. However, the extrapolation is still performed linearly and a correction is then introduced for the Forbes effect [432]. The correction can be calculated (it lies in the range 1-3%) from the known spectral distribution of the transparency of the atmosphere during the calibration process.

The calibration is performed in a dry, mountainous region with good atmospheric transparency and stable optical properties of the atmosphere at certain periods of the year. The stability of the optical properties is monitored against the solar aureole. The optical properties of the atmosphere are assumed constant during the calibration process, with the relative solar aureole per unit air mass varying nonmonotonically by not more than 10%.

This method of calibration is based on the laws of atmospheric optics [371-376,433,434]. The light beam passing through the Earth's atmosphere is attenuated by scattering and absorption. According to Bouguer's law, the ratio of the spectral flux density of solar radiation E_λ on the Earth's surface to the corresponding flux density $E_{0\lambda}$ outside the atmosphere is given by (4.4).

The absolute air mass is a function of not only the position of the Sun above the horizon, but also the height of the locality above sea level. This is taken into account in (4.1) by introducing the factor p/p_0. When the Sun is low above the horizon, the results must be corrected for the distortion of the optical ray paths due to the curvature of the Earth and the spherical shape of the atmosphere, as well as the curvature of the solar rays due to the gradual and continuous variation of the refractive index with height. Accurate values of the air mass are tabulated in [371] and [373].

Let P_λ in (4.4) represent the factor $\exp(-\alpha_\lambda)$. Bouguer's law for the atmosphere then assumes the form

$$E_\lambda/E_{0\lambda} = P_\lambda^m. \qquad (4.5)$$

The quantity P_λ is called the spectral transparency coefficient of the Earth's atmosphere. In view of (4.5), the short-circuit current of a standard cell on the Earth's surface can be written in the form

$$I_{sc} = \mathscr{E}F \int_0^\infty E_{0\lambda} P_\lambda^m S_\lambda \, d\lambda, \qquad (4.6)$$

where S_λ is the spectral sensitivity, F is the area of the standard cell, and $\mathscr{E}$ represents the ellipticity of the Earth's orbit and is equal to the ratio of the solar flux density at the given point on the orbit to the solar constant.

Since the short-circuit current of the standard cell under AM0 conditions is given by

$$I_{AM0} = F \int_0^\infty E_{0\lambda} S_\lambda \, d\lambda, \qquad (4.7)$$

we have from (4.6) and (4.7)

$$I_{sc} = \mathscr{E}I_{AM0} \int_0^\infty E_{0\lambda} P_\lambda^m S_\lambda \, d\lambda \Big/ \int_0^\infty E_{0\lambda} S_\lambda \, d\lambda. \qquad (4.8)$$

The ratio of the integrals in (4.8) can be looked upon as the integrated atmospheric transmission coefficient:

$$\int_0^\infty \frac{E_{0\lambda} P_\lambda^m S_\lambda \, d\lambda}{E_{0\lambda} S_\lambda \, d\lambda} = P_{eff}^m. \qquad (4.9)$$

Substituting (4.9) in (4.8) and taking the logarithm of both sides, we obtain

$$\log \mathscr{E}I_{AM0} = \log I_{sc} - m \log P_{eff}. \qquad (4.10)$$

This equation is used to determine I_{AM0} by graphical extrapolation. The correction ψ [432] for the Forbes effect [434] is then applied to this current. It is found from the effective transparency P_{eff} at the beginning (absolute air mass m_1) and end (m_2) of the measurements under stable atmospheric parameters:

$$\psi=1+0{,}384\,(m_1{}^2+m_2{}^2+4m_1m_2)\,(\log P_{\mathrm{eff}}(m_2)-\log P_{\mathrm{eff}}(m_1))/(m_2-m_1).\,(4.11)$$

The effective transparency coefficient P_{eff} of the atmosphere can be determined as follows. For monochromatic radiation, (4.10) shows that

$$\log I_\lambda=\log \mathscr{E}I_{\lambda\,\mathrm{AM0}}+m\log P_\lambda. \qquad (4.12)$$

In this case, P_λ is independent of m. It is clear from the last equation that the spectral transparency P_λ of the atmosphere is equal to the antilog of the slope of the straight line giving log I_λ as a function of the air mass. Thus, by measuring the short-circuit current of the standard cell under illumination by narrow and practically monochromatic radiation as a function of the air mass, we obtain the spectral transparency for the given wavelength. Having obtained P_λ for the solar range in which the given standard cell is sensitive, we can estimate the effective transparency coefficient P_{eff} of the atmosphere from (4.9).

Standard solar cells have been continuously calibrated in the USSR since 1965. This is done regularly about once or twice a year [394,395,418,419,430-433] in the neighborhood of Alma-Ata at the High-Altitude Station of the Shternberg Astronomical Institute (43°N, 77°E; 3014 m above sea level) by a group headed by E. V. Kononovich, which has developed a computer program for calculating I_{sc} and has equipped a solar telescope for these measurements. At the All-Union Scientific Research Institute for Current Sources, V. Ya. Kovalskii et al. have developed the equipment and methodology for high-altitude calibrations.

The relative air mass M is a function of the true time τ and depends on the declination δ of the Sun on the given day and the latitude φ of the locality:

$$M=\operatorname{cosec}\theta=\sin\varphi\sin\delta+\cos\varphi\cos\delta\cos\tau,$$

where θ is the altitude of the Sun above the horizon.

The time at which the measurements are taken in the course of calibration is measured on the legal scale T_l. The true time is then determined from the formula

$$\tau = T_l - h - \Pi + D - \Delta\tau - 12,$$

where Π is the time zone of the point of observation, D is the longitude, $\Delta\tau$ is the equation of time (difference between the mean and true time), and h is the difference between the zone and legal time.

Calculations are then made to determine the relative air mass as a function of legal time.

The relative air mass is determined for each measurement of the short-circuit current of the photosensitive cell in the aureole photometer I_{au} (pointing at the aureole around the Sun) and the corresponding current I_s (photometer pointing directly at the Sun). The ratio $R_s = I_{au}/I_s m$ is then evaluated. This is referred to as the relative solar aureole.

The stability of the relative solar aureole is determined for half the solar day, and the range of air masses for which R_s varies non-monotonically to within less than 10% is determined. This yields the initial and final values m_1 and m_2 of the air mass range for the stable atmosphere.

An example of a determination of I_{AM0} by this method is illustrated in Fig. 4.15 for two standard cells.

A graph of log $I_\lambda = f(m)$, where I_λ is the short-circuit current of the filter monochromator, is constructed in order to determine the Forbes correction. The slope κ of this line gives the spectral transparency P_λ of the atmosphere:

$$\tan \kappa = -\log P_\lambda.$$

The complete spectral distribution of atmospheric transparency can be determined by calculating P_λ for all the filters.

The effective transparency P_{eff} of the atmosphere is calculated for the initial (m_1) and final (m_2) values of the air mass interval of the stable atmosphere using (4.9), and the Forbes correction ψ is calculated from (4.11).

The final result of the calibration (the current I_{AM0_T}) is calculated from the resulting current I_{AM0} under extra-atmospheric conditions which, in turn, is found by extrapolation to zero air mass after introducing the Forbes correction and the correction for the ellipticity of the Earth's orbit:

$$I_{AM0_T} = I_{m=0}\psi/\mathscr{E}.$$

The ellipticity coefficient of the orbit is calculated from the

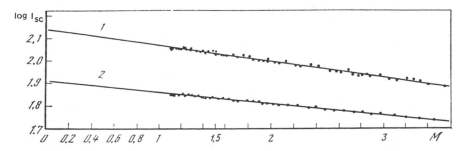

FIG. 4.15. Measured short-circuit current of standard solar cells
plotted against the air mass on a semilogarithmic scale. The experimental data were obtained as a result of high-altitude measurements on 26 June 1982, near Alma-Ata: 1) silicon; 2) AlGaAs–
GaAs cell.

ratio of the flux density at the top of the atmosphere for a given
position of the Earth in orbit to the solar constant, i.e., the energy
flux density at the mean Earth–Sun distance. The coefficient $\mathscr{E}$
varies from 0.965 to 1.035 in the course of the year. The value of $\mathscr{E}$
for any particular day can be looked up in the Astronomical Yearbook
[435].

Determination of the extra-atmospheric current
for standard cells, using the spectral transparency
of the atmosphere measured at high altitudes

A method of calibration, which may be used under terrestrial
conditions, consists of the simultaneous single determination of the
solar cell current and the spectral distribution of the solar radiation.
The extra-atmospheric short-circuit current is calculated from the
formula proposed in [436]:

$$I_{\text{AM0}} = I_{\text{AMT}} \int_{0}^{\infty} E_{\text{AM0}}(\lambda)\, S'\, d\lambda \left/ \int_{0}^{\infty} E_{\text{AMT}}(\lambda)\, S'(\lambda)\, d\lambda \right.,$$

where I_{AMT} is the standard-cell current measured under terrestrial
conditions, $E_{\text{AM0}}(\lambda)$, $E_{\text{AMT}}(\lambda)$ are the spectral energy densities
of extra-atmospheric and terrestrial solar radiation, respectively,
and S' is the relative spectral sensitivity of the standard cell.

This method of calibration relies on the accurate determination of the spectral energy distribution of terrestrial solar radiation (which is quite a problem) and the relative spectral sensitivity of the cell.

The calibration of standard cells, including measurement of the time-dependent spectral distribution of terrestrial solar radiation, can be substantially accelerated and simplified by using fast-filter, prism, or diffraction-grating monochromators-spectrometers.

A rapidly scanning spectroradiometer has been developed for the wavelength range 0.25-2.5 μm with a transmission band of 100 Å throughout the range and scanning rate up to 0.1 μm/sec [437]. The dispersing element of the spectroradiometer is a diffraction grating. The wavelength is automatically varied and recorded. A light-scattering element and an optical sensor are mounted at entry to this instrument so that the direct and scattered components of the incident solar flux can be determined separately. The integrated photocurrent delivered by thin-film copper sulfide—cadmium sulfide heterojunction solar cells was determined for different solar flux densities with the spectrum being recorded at the same time. This was necessary for the comparison between field and laboratory calibrations of solar cells.

The methodology and results of high-altitude measurements of the extra-atmospheric parameters of silicon and AlGaAs—GaAs and Cu_2S—CdS solar cells are described in [418,419,438].

Calibration of terrestrial standard cells,
using high-altitude measurements

If the relative air mass obtained during high-altitude measurements is converted into absolute air mass, the logarithm of the short-circuit current as a function of the absolute air mass (see Fig. 4.15) can be used to determine the short-circuit current of standard cells not only for AM0, but also for AM1, AM1.5, and AM2, as well as other large air masses.

However, when standard cells are calibrated for the international standard solar spectrum, the spectrum of terrestrial solar radiation used in measurements must be matched to this spectrum not only in respect to the air mass but also all other parameters such as flux density, the amounts of water vapor and ozone, the turbidity coefficient, and the selectivity index. By determining the solar spectrum on the day the tests are performed under high-altitude conditions, and then comparing it with the standard spectrum, we can introduce the necessary correction to the current produced by a standard cell, using a graph such as that shown in Fig. 4.15, for any value of the absolute air mass. Thus, high-altitude measurements provide us with a basis for obtaining relatively accurate calibration values for the current produced by a standard cell, which can then be used to estimate the parameters of terrestrial solar cells. The reduction to the standard spectrum can also be performed without studying the solar spectrum on a particular day in detail: it is sufficient to know the depth of a few characteristic bands in the spectrum in order to estimate the concentration of water vapor, ozone, and aerosols on that day [431].

The above method was used at the High-Altitude Station of the Shternberg Astronimical Institute near Alma-Ata on 3 and 4 August 1978. Terrestrial calibration of silicon, gallium arsenid, and cadmium sulfide standard cells was performed [418]. Two other methods were also employed.

The flux density of terrestrial solar radiation was measured with the AT-50 thermoelectric actinometer. The amount of precipitated water vapor was determined with a filter monochromator, using the transmission ratio at 1.1 and 0.84 μm. The turbidity of the atmosphere was estimated from transmission at 0.56 μm and the intensity of the solar aureole measured with an aureole photometer. Ozone absorption was determined from the mean data for August at the latitude of Alma-Ata.

The short-circuit current of standard solar cells was measured for relative air masses between 1.5 and 2.5. The optical air mass under these conditions is practically the same for all the components of the atmosphere, and Bemporad corrections are small, so that the absolute air mass can be determined from the actual atmospheric pressure, using (4.1).

The absolute air mass of 1.5 corresponded on different days to relative air mass of 2.1-2.2. The other parameters were as follows: on 3 August, 1978, when the altitude of the Sun above the horizon corresponded to an absolute air mass of 1.5, the flux density of direct solar radiation was $E_d = 827$ W/m^2, the thickness of the precipitated water layer was $\omega = 2.41$ cm, the turbidity coefficient was

$\beta = 0.048$, the ratio of the intensity of solar aureole to the intensity of the Sun was $R_s = 0.09$; on 4 August 1978, when the absolute air mass was 1.5, the corresponding values were $E_d = 904$ W/m^2, $\omega = 1.47$ cm, $\beta = 0.017$, and $R_s = 0.07$.

The relative dependence of the short-circuit current of standard cells on the concentration of water vapor was then calculated for different turbidity coefficients. The values of ω and β, measured on a particular day, were used to calculate the correction due to the difference between the values of ω and β at the time of measurement and the corresponding values for the standard international spectrum of terrestrial solar radiation. The correction for the difference between the radiation flux density at the time of measurement and the standard values was introduced by simple multiplication of the current by the necessary factor.

This method was found to produce calibration short-circuit currents of terrestrial standard solar cells that were in good agreement with values obtained by other methods [366,418,419,439].

Determination of the short-circuit current of standard cells under standard extra-atmospheric and terrestrial solar spectra from measured spectral sensitivity

The spectral sensitivity of solar cells is one of the basic parameters that can be used in semiconductor photoelectric devices [6,7, 15,16] to estimate the efficiency at all stages of conversion into electric power. In particular, the spectral sensitivity can be used in choosing the optimum semiconductor material for a solar cell [82], to estimate the influence of electrostatic drag fields [80,84,419], to determine the recombination parameters [85-90], to determine the diffusion length of minority carriers in the base layer of cells [91-93], including the nonuniform distribution of defects with depth in the base layer [97], and to obtain the rise in current due to the deposition of antireflective coatings [290,23].

Measurements of the spectral sensitivity performed with conventional monochromators under low levels of illumination (between 10 and 20 mW/cm^2) cannot be used when the solar cell has a nonlinear dependence of current on illumination because, even when concentration is not employed, the flux density incident on the surface of the cell is greater by several orders of magnitude, i.e., between 85 mW/cm^2 (AM1.5) and 136 mW/cm^2 (AM0). Moreover, the diffusion length L_b of minority carriers in the base layer is a function of the level of illumination, since an increase in the concentration of carriers injected by light produces an initial sharp rise in L_b which thereafter remains practically constant [22,92,93]. At low levels of illumination, the deviation from linearity is due to recombination processes, whereas under ultrahigh levels it is due to power losses in the spreading resistance.

It is quite clear that measurements of the spectral sensitivity of standard solar cells (with a view to subsequent conversion to the standard spectral energy distribution and determination of the calibration photocurrent) must be performed for levels of illumination that are close to the conditions encountered when standard cells are used in practice [440-447].

One of the first systems used to measure the spectral sensitivity under illumination similar to the incident solar flux density consisted of a powerful hot-filament halogen-tungsten lamp and eight narrow-band interference filters, calibrated for radiation flux density [440]. Two other similar systems (eight filters with full width at half height of the transmission band of about 200 Å, and eighteen filters covering the range between 0.35 and 1.2 μm) were subsequently used at the Lewis Research Center (Cleveland, Ohio) [442]. The source of radiation was again a halogen lamp (1000 W output). The results were used to convert the wavelength dependence of the short-circuit current to the standard solar radiation spectrum and to compare the calculated photocurrent with calibrations performed at high altitudes, on rockets, and on sounding balloons [426-428].

The filter monochromator for measuring the spectral sensitivity of solar cells was substantially improved in [443]. The source of radiation in front of the narrow-band filters was in the form of a pulsed lamp producing 600 J per pulse (an aluminum reflector was placed behind the lamp). However, the necessary flux uniformity was not achieved (the variation was ±8% across the field). The short light pulse from the xenon flash tube ensured that neither the solar cells under examination nor the interference filters were overheated. The system can be used for rapid determinations of the spectral sensitivity, and absolute photocurrents are obtained by comparing the measured short-circuit current with the current of a standard cell. The optical system used to measure the spectral sensitivity

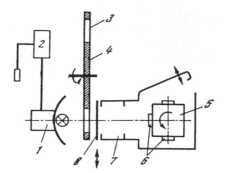

FIG. 4.16. Optical system of the apparatus used to measure the spectral sensitivity of solar cells under high level of illumination by a flash lamp working with interference filters: 1) xenon flash lamp and reflector; 2) supplies for the lamp; 3) interference filters; 4) rotating disk (filter-holder); 5) rotating specimen-holder; 6) solar cell; 7) container; 8) shutter used to cut off scattered light in intervals between measurements.

of solar cells is shown in Fig. 4.16. Light from the flash lamp 1 is sent onto the solar cell through one of sixteen interference light filters mounted on the rotating disk-holder 5.

The pulsed current produced by the solar cells is measured by an electronic circuit and is displayed in digital form. The flux density produced by the flash lamp without the light filters in position exceeds 50 solar constants, which means that the sensitivity can be measured under near-working conditions.

Absolute calibration of the system is performed against a standard cell whose sensitivity is measured on a monochromator calibrated with a nonselective thermoelectric detector at 0.546 μm. The calibration uncertainty is ±2% (absolute) and ±1% (relative).

When lasers or high-intensity light sources (hot-filament lamps) [440-442] and flash lamps [443] working with interference filters are used in spectral measurements, the resulting distribution of charge carriers within the solar cell is not the same as that produced by extra-atmospheric radiation. This means that the most reliable data on the sensitivity of solar cells are obtained under the simultaneous illumination of solar cells by a modulated beam of monochromatic radiation and an unmodulated beam simulating solar

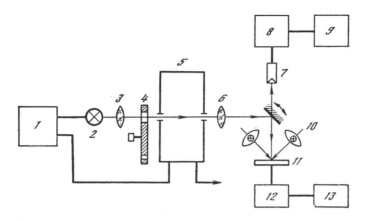

FIG. 4.17. Apparatus used to measure the spectral sensitivity of standard solar cells: 1) supplies for lamp and monochromator; 2) lamp used to illuminate the entrance slit of the monochromator; 3) condenser; 4) modulator; 5) monochromator; 6) focusing lens; 7) thermoelectric radiometer; 8) amplifier for the output voltage of the thermoelectric radiometer; 9) graph-plotter; 10) lamps simulating solar radiation; 11) thermostated standard solar cells; 12) tuned amplifier for the current produced by the standard solar cell; 13) recording equipment.

radiation with the necessary spectrum and flux density. In the first attempts to apply this method to the calibration of standard solar cells [436,446], the necessary level of carrier injection was achieved with hot-filament lamps. However, the spectrum of the illuminating radiation must reproduce the solar spectrum, since the nonlinearity of the spectral characteristic increases rapidly with increasing wavelength [444,445,447].

This method of calibration has been substantially developed and improved [447,448]. Uncertainties due to the nonlinearity of the spectral characteristic and the mismatch between the distribution of carriers generated by light and the true distribution were reduced [448] by using nonselective radiometers to measure monochromatic radiation, high-luminosity monochromators, better simulation of the solar spectrum, and a modulator in which the higher harmonics were suppressed as much as possible. The illuminating radiation was produced by halogen lamps incorporating built-in interference filters, capable of reproducing the solar spectrum [407], and a special hollow thermoelectric radiometer with a substitution coil [105] was used in the absolute calibration of the monochromator. Calibration against a black body is also useful (in addition to the built-in substitution coil). The nonselective radiation detector can also be in the form of a thin-film sensor incorporating multicomponent thermoelectric layers [103], a small device consisting of a large number of metal thermocouples [104], an evacuated thermopile [449], or a strip film detector with deposited thermocouples [450].

Figure 4.17 illustrates a system used to measure the spectral sensitivity. It was specially developed for the calibration of standard solar cells.

The particular feature of this system is the presence of the headlight-type lamps in which the reflector and transmitting window carry multilayer interference filters that correct the spectrum of the KGM-6.6-200 lamp to bring it closer to the solar spectrum [407]. An irradiance of 1360 W/m^2 is produced on the surface of the cell and is monitored by a thermoelectric radiometer with a large field of view. The radiometer is carefully calibrated in a broad spectral range. The illuminating lamps are supplied by stabilized sources with minimum high-harmonic content.

Monochromatic radiation of sufficient intensity is produced by using the MDR-3 diffraction grating monochromator (600 lines/mm). Higher-order spectra are excluded by a device (variable quenching resistance in the lamp circuit, coupled to the mechanism rotating the diffraction grating in the monochromator), which reduces the color temperature of the lamp by reducing the lamp current whenever the long-wave part of the spectrum is being examined. The short-circuit current under monochromatic illumination is measured at different points on the photoactive surface of the standard solar cell and is then averaged over the entire working area.

The monochromatic flux modulated at 900 Hz is directed onto the cell. The mutual disposition of the monochromator cell and the modulator and the shape of the modulator window are chosen so that the monochromatic modulated beam is as nearly sinusoidal as possible. Since the measurement has to be performed under short-circuit conditions, the alternating signal is recorded through a dividing capacitor and the solar cell is shunted by a resistance of the order of 0.5 Ω. The high-frequency component of the short-circuit current is fed into a calibrated tuned amplifier whose output is converted into a proportional signal and is recorded in digital and graphic form. The data can also be printed out on punched tape for use in a computer.

When the spectral sensitivity of several silicon solar cells [448] was used to calculate the extra-atmospheric solar spectrum [358], and this was followed by a calculation of the integrated short-circuit current of the cells, it was found that, in the case of nonlinear solar cells, the uncertainty in the calibration current for extra-atmospheric conditions, due to the fact that the sensitivity was measured without using a Sun simulator, could reach up to 7%.

Comparative conversion of the spectral sensitivity of silicon solar cells [448] to the extra-atmospheric solar spectrum of Makarova and Kharitonov [358], Thekaekara [356,364], and Johnson [369] has confirmed the conclusions reported in [370], namely, that the short-circuit currents calculated from the Makarova—Kharitonov spectrum are higher by 3-3.5% than those determined from the Thekaekara spectrum, but are close to those calculated from the Johnson spectrum (despite the appreciable discrepancy between the solar constants), since the first and last spectra are close to each other in the spectral sensitivity range of high-grade silicon cells (between 0.3 and 1.1 μm).

There is one further method of measuring the spectral sensitivity of standard solar cells that ensures low uncertainty in subsequent calculations of the integrated photocurrent under illumination by the standard spectrum [102,395].

The short-circuit current of a standard cell (under both extra-atmospheric and standard terrestrial spectra) can be written in two forms, namely,

$$I_{sc} = F_s \int_0^\infty E_s S_s \, d\lambda = F_s g_m \int_0^\infty E_s S'_s \, d\lambda,$$

where F_s is the photosensitive area of the standard cell, E_s is the flux density of the standard (extra-atmospheric or terrestrial) solar radiation, S_s and S'_s are the absolute and relative spectral sensitivities of the standard cell, respectively, and g_m is a scaling factor.

The second relation can be used to calculate the absolute spectral sensitivity and the standard value of the integrated short-circuit current when the cell is illuminated by a source with any spectrum, provided the relative spectral sensitivity has been measured with the usual monochromator producing low monochromatic flux density, and the absolute value of the scaling factor has been determined for high radiation fluxes, i.e., those corresponding to the solar flux.

The scaling factor can be determined in two ways: by measuring the integrated current through the cell when it is illuminated by a standard lamp with a known color temperature, or by calibrating the relative spectral measurements in some narrow spectral interval against a standard thermocouple [395].

This method of measuring and calibrating the spectral sensitivity of solar cells was developed further and improved at the All-Union Scientific Research Institute for Opticophysical Measurements (VNIIOFI) where, in 1978, the first high-precision system was built for measuring the relative spectral sensitivity of radiation detectors in the wavelength range 0.35-2.5 μm, and for measuring the relative spectral flux density due to sources in the wavelength range 0.4-1.0 μm [451]. The Institute also has a system for measuring the integrated sensitivity of linear detectors and for measuring the deviation from linearity when the nonlinearity increases with increasing signal strength [452]. When the scaling factor is measured to obtain the absolute sensitivity, a determination is made of the detector signal in the ultraviolet due to a standard source of ultraviolet radiation at 0.254 μm (a group of DRB-8 low-presssure quartz lamps [453, 454]), whereas at wavelengths between 0.3 and 2.5 μm the sources are SI-10-300U hot-strip tungsten lamps [455].

It is important to note that the reduction of the relative spectral sensitivity by the scaling factor leads to a proportional increase in the sensitivity of the solar cells (measured under low illumination) at all wavelengths, while high levels of illumination and changes in the spectral composition of the radiation usually have a different effect on the short- and long-wave parts of the spectral sensitivity.

Because of this, measurements of the spectral sensitivity of low-grade silicon solar cells with nonlinear characteristics by different methods yield different results (curves 1-3 in Fig. 4.18). At the same time, in the case of high-efficiency and sufficiently linear silicon solar cells, the principal methods of measuring the absolute

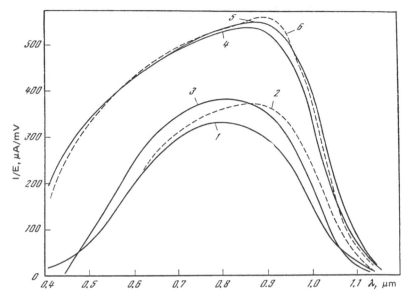

FIG. 4.18. Absolute spectral sensitivity of silicon solar cells with a deep p—n junction and without an antireflective coating (doped layer thickness $l \geqslant 1.2 \ \mu m$) (1-3) and with an antireflective coating and a shallow p—n junction ($l \leqslant 0.3 \ \mu m$) (4-6): 1, 4) measurements on a monochromator under low levels of illumination using a calibrated thermocouple; 2, 6) same as 1, 4 but with a modulated flux simulating solar radiation; 3, 5) measurements 1, 4 recalculated with the aid of the scaling factor determined from measurements under a standard lamp and high levels of illumination.

spectral sensitivity do yield similar results (curves 5 and 6 in Fig. 4.18) which can be used with confidence in the subsequent calculation of the integrated short-circuit current of standard solar cells under illumination by a source with any standard spectrum. Illumination by simulated solar radiation [448] produces a higher long-wave spectral sensitivity (determined by the base layer; see curves 2 and 6 in Fig. 4.18), whereas measurements using the scaling factor [395,102,451-455] yield higher sensitivity in practically all spectral ranges (curves 3 and 5 in Fig. 4.18) as compared with the spectral sensitivity obtained with a monochromator under low illumination (curves 1 and 4 in Fig. 4.18). However, the most reliable results (curves 5 and 6 in Fig. 4.18) are so close to one another that we may conclude that the agreement between these measurements is good.

Solar cells incorporating new semiconducting materials such as, for example, gallium arsenide with a top window in the form of a solid solution of aluminum in gallium arsenide, which forms a heterojunction with gallium arsenide, or thin-film copper sulfide—cadmium sulfide heterosystems, do not exhibit the above dependence of spectral sensitivity on the incident spectral composition or flux density. This suggests that their photoelectric characteristics are linear for flux densities in the range between 0.1 and 136 mW/cm^2. The important features here are the high absorption coefficient of these materials, the small diffusion length of minority carriers, which depends on the parameters of the original material, and the lower carrier injection level as compared with silicon. The fact that the characteristics of cells made from these new semi-conducting materials are linear facilitates the calibration of the corresponding standard solar cells.

Determination of the metrological parameters of solar cells and batteries made from different semiconducting materials by direct measurement under extra-atmospheric conditions

Since it is quite difficult and expensive to calibrate standard solar cells in space, and then recover them, attempts have been made to measure the parameters of silicon and gallium arsenide solar cells in space by using a telemetry system. The data can then be assigned with a high degree of confidence to duplicates of these cells that remain on Earth and have the same spectral and current-voltage characteristics. Such experiments were performed on the automatic interplanetary stations in the Venera series [456]. The solar cell panels carried by these stations have a number of reliable telemetry channels, including channels for measuring the temperature of the batteries. The batteries are accurately aligned with the Sun, so that their characteristics in space can be reliably estimated.

The electrical parameters of solar cells and modules tested on the Venera stations were measured at all stages of fabrication and under illumination by high-grade Sun simulators in the laboratory. The latter measurements were performed on a simulator incorporating a xenon lamp and interference filters, similar to that described in [21], and on a simulator with a metal-halogen (gas-discharge) lamp [400-402]. Measurements were also made under the S-1 simulator incorporating filament halogen lamps with colored glass filters [395] or with only a heat-absorbing layer of distilled water [404]. The parameters of batteries were measured in the course of fabrication and assembly under hot-filament lamps without spectral correction [403] and headlight-type lamps with spectral correction [407].

High-grade simulators were adjusted under laboratory conditions against standard solar cells of the necessary type. The standards were calibrated for the flux density and spectrum of extra-atmospheric solar radiation by extrapolating measurements of the short-circuit current under natural terrestrial solar radiation to zero air mass. The precision of calibration was checked by comparing the resulting parameters with data obtained by similar solar cells used as sensors on different spacecraft. The extra-atmospheric parameters were also calculated from the absolute spectral sensitivity.

The prototype batteries mounted on the Venera stations consisted of four individual generators, 0.5 m^2 each. The generators were assembled from modules consisting of solar cells of the following types: silicon cells with deep p—n junctions (1.0-1.2 μm), silicon cells with relatively shallow junctions (0.3-0.5 μm), and AlGaAs—GaAs cells. All the silicon modules consisted of parallel-connected, large cells, transparent to the infrared. The cells were protected by a common glass cover. A highly reflecting metal grid, lying accurately above the cell connections, was deposited on the rear surface of the glass. The design of these modules was successfully tested on Venera-9 and Venera-10 [142]. The combination of the transparent solar cell and the reflecting grid ensured that the equilibrium temperature of the batteries in space could be reduced to about 45°C. Modules of the AlGaAs—GaAs cells were opaque and did not have the heat-reflecting grid, so that their equilibrium working temperature was 65-70°C.

Sufficiently accurate duplicates of generators and of the solar cells from which the generators were assembled were retained in the laboratory until the end of the mission. Standard cells were then selected from them. The selection and calibration of these standard cells was carried out according to the extent to which

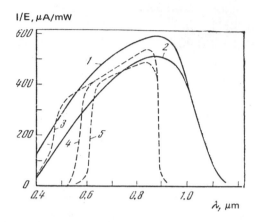

FIG. 4.19. Absolute spectral sensitivity of solar cells mounted on Venera-13 and Venera-14: 1) silicon cell with p—n junction depth 0.3-0.5 μm; 2) silicon cell with junction depth 1.0-1.2 μm; 3) p-AlGaAs—p-GaAs—n-GaAs heterostructure with a thin layer (less than 1 μm) of the solid solution p-AlGaAs; 4, 5) the same heterostructure with a p-AlGaAs layer of 10 and 15 μm, respectively.

there was agreement between laboratory, high-altitude, and space calibrations.

For Venera-13, the generator consisted of AlGaAs—GaAs solar cells with spectral characteristics similar to curve 5 in Fig. 4.19, whereas for Venera-14 the characteristic was similar to curve 4. The solar distribution derived by Makarova and Kharitonov [358] was used in all calculations on the spectral distribution.

Table 4.2 lists the measured electrical parameters of a number of prototype cells, obtained under a Sun simulator and in space. The simulator incorporated hot-filament lamps without spectrum correction and was adjusted in all cases against a standard silicon solar cell with a deep p—n junction and highly stable sensitivity (checked over a period of more than ten years). The short-circuit current and temperature were measured at the beginning of the mission, when the solar flux density on the cell surface was 1.01 of the solar constant. The conversion factor, which was applied to simulator measurements to obtain the extra-atmospheric data, was determined as the ratio of the load current generated by the battery

TABLE 4.2

Metrological Parameters of Solar Cells Measured under a Sun Simulator
and from Venera-13 and Venera-14 Automatic Interplanetary Stations

Solar cell	Load current, A		t, °C	Conversion factor	
	under simulator	in flight	in flight	experiment	calculation
Venera-13					
Silicon solar cell with deep p—n junction	1.71	1.74	45	1.01	1.00
	2.10	2.10	45	0.99	1.00
AlGaAs—GaAs (AlGaAs thickness 15 μm)	1.47	1.65	65	1.12	1.11
Venera-14					
Silicon solar cell with shallow p—n junction	1.77	2.00	45	1.12	1.13
	1.09	1.20	45	1.09	1.13
AlGaAs—GaAs (AlGaAs thickness 10 μm)	1.42	1.70	70	1.20	1.21

Note. Temperature during measurements under simulator 50°C.

in space to the load current obtained under the simulator. The theoretical value of this factor was calculated from the spectral sensitivity of the solar cells.

Thus, if the hot-filament simulator without spectrum correction is adjusted against the standard silicon cell with a deep p—n junction, the load current delivered by a battery consisting of these cells in space must be the same as under the simulator. For batteries assembled from silicon cells with shallow p—n junctions, the current is higher by an average of 10% as compared with the current under the simulator. Under the same conditions, the current produced by a battery consisting of AlGaAs—GaAs cells is higher by l0-20% (depending on the spectral characteristics).

This is due to the considerable difference between the spectral sensitivity of AlGaAs—GaAs cells and silicon solar cells with a shallow p—n junction, on the one hand, and the sensitivity of silicon cells with a deep p—n junction (as in the standard), on the other. For a given short-circuit current due to a hot-filament lamp (color temperature about 2800 K), cells fabricated by different technologies from different semiconducting materials will have different short-circuit currents under illumination by solar radiation outside the Earth's atmosphere (color temperature about 6000 K). As the depth of the p—n junction in silicon solar cells is reduced, their spectral sensitivity changes mostly in the short-wave part (see Fig. 4.19), where the solar intensity is much higher than the intensity of the hot-filament lamp. This means that the current recorded under extra-atmospheric conditions will be greater than the current recorded under a simulator. For p-AlGaAs—p-GaAs—n-GaAs cells, a reduction in the thickness of the AlGaAs optical filter-window from 10 μm to less than 1.0 μm will extend the spectral sensitivity range into the short-wave region (see Fig. 4.19, curves 3-5), and the extra-atmospheric parameters will increase as compared with the parameters determined under a simulator adjusted against a silicon standard. The short-circuit current calculated from the spectral sensitivity shows that, for the heterostructure cell with a small filter-window thickness, the conversion factor is much higher and amounts to 1.54.

The introduction of the necessary corrections during the simulator adjustment does not always lead to the complete removal of uncertainties because the conditions of measurement (radiation parameters) and the characteristics of the cells themselves do not remain constant in all cases. This means that, if a precise prediction of the working parameters of batteries in space is to be made, the measurements must be made with standards that not only employ solar cells of the same type, but the cells must be fabricated by the same technology.

The simulator can be adjusted against stable standard cells fabricated by a technology different from that of the solar cells to be examined, but this requires the introduction of correction factors determined in careful laboratory and field tests.

4.5   Measurement of the parameters of solar cells and batteries

The principal descriptor of solar cells and batteries is the load current-voltage characteristic. It can be used to determine the electric power generated by them in the form of the product $I_{opt}U_{opt}$, to estimate the contribution of the gap potential to the open-circuit voltage, to determine optical and photoelectric losses from the short-circuit current and the fill-factor of the current-voltage characteristic, and to calculate the efficiency of conversion of solar radiation into electric power from the ratio of the power generated by the cells and batteries to the incident solar power, which can be measured with a calibrated standard solar cell.

The short-circuit current is determined by converting the measured absolute spectral sensitivity to the standard extra-atmospheric or terrestrial solar spectrum. The quality of solar cells and batteries and the number of faulty cells in a battery can also be estimated indirectly by measuring the forward and reverse branches of the dark current-voltage characteristic, the integrated absorption coefficient for solar radiation of the surface of the battery (calculated from the measured spectral reflection coefficient), and the integrated thermal emission coefficient of the battery surface, which is different for faulty and satisfactory cells.

The above methods can be used to measure the parameters of solar cells and batteries under laboratory, field terrestrial, and space conditions.

We shall now consider some methodological problems involving quality control, the determination of the parameters of solar cells and batteries made from different semiconducting materials for different practical applications, and uncertainties in measurement and prediction of the working cell parameters.

Measurement of terrestrial and extra-atmospheric parameters of solar cells under laboratory conditions

In June 1982, a simulator of terrestrial solar radiation was used in Budapest to perform joint Soviet-Hungarian measurements of the current-voltage characteristic and efficiency of solar cells under laboratory conditions [423].

The source of light in the simulator was a high-pressure xenon lamp whose spectrum was corrected by an interference filter. The simulator was adjusted against the PS-9 standard solar cell (sensitive area 30 X 35 mm), developed and calibrated in the USSR. As already noted, this cell has been put forward as a standard for countries belonging to the Socialist Economic Community. The calibration was performed for terrestrial AM1 conditions (direct solar flux density 1000 W/m^2) and AM1.5 conditions (850 W/m^2). Measurements under the simulator were made with an instrument (developed in Hungary) for the automatic determination and recording of the current-voltage characteristic, which was connected to a minicomputer, so that the optimum parameters of solar cells could be determined simultaneously. The instrument makes use of a four-probe method of measuring the current, with a separate circuit for connecting a voltmeter, which is much more accurate than the two-probe system (Fig. 1.18) for measuring the voltage across the solar cell. Since the current flowing in the voltmeter circuit in the four-probe system is very small, the voltage drop across the junction resistance between the solar cell contact and probe and across the lead is negligible, so that the voltmeter measures the voltage directly across the solar cell. Experiments have shown that, for cells of 5.4 cm², standard radiation flux density, and short-circuit current of 160 mA, the efficiency measured by the two methods is the same; for an area of 10.5 cm² and short-circuit current of 300 mA, the efficiency measured by the two-probe system is 12.1 instead of 14.1% indicated by the four-probe scheme. When the cell area is 24 cm² and the short-circuit current is 670 mA, the two efficiencies are 8.1 and 11.3%, respectively.

Whatever the electrical circuit used, an increase in the resistance between the cell contact and the current probe, the resistance of the leads, and the internal resistance of the ammeter will ensure that the solar cell parameters will be measured on the part of the current-voltage characteristic that is more distant from the short-circuit current, so that the uncertainty in the measurement will be very appreciable for cells with high series resistance.

In precise measurements of short-circuit currents, the reverse voltage can be balanced by an additional external voltage source. This is particularly convenient in measurements under higher solar flux densities or when the solar cell has a large photosensitive area. For example, this is used to measure the electrical parameters of cell blocks (modules with parallel-connected solar cells of large dimensions, which produce high currents for small voltages [141, 142]).

The Sun simulator measurements performed in June 1982 in Hungary were compared with data obtained under the same simulator when it was adjusted against a solar cell that had been compared with the standard adopted in the USA (the standard solar cell used in the USA for AM1 measurements on terrestrial cells was demonstrated during the Soviet-American seminar in 1977

[393]). When adjustment was made against the US standard, the efficiency of the cells was found to be higher by an average of 8% for the Hungarian cells and 6% for the Soviet cells. This appears to be due to differences between the methods used in the USSR and USA for standard cell calibration.

The Soviet standard cells were calibrated against natural solar radiation in the region of Alma-Ata (3000 m above sea level) and by calculation from the absolute spectral sensitivity (see Section 4.4). In the latter case, the calibration current was determined by converting the absolute sensitivity of standard cells to the standard AM1 spectrum with a narrower water vapor absorption band in the region of 0.9 μm than that adopted in [376,387,389] (a more accurate band shape was obtained after the publication of these papers and this was taken into account in the AM1.5 international spectrum [391]). Figure 4.20 shows the standard terrestrial spectra between 0.8 and 1.1 μm under AM1 [387], AM2 [376,389], and AM1.5 [391] conditions. It is clear that the absorption band due to water vapor in the region of 0.9 μm is shown by modern data [391] to be somewhat narrower (curve 2). This part of the spectrum corresponds to the maximum spectral sensitivity of silicon solar cells and, if the US standard cell was calibrated against the AM1 spectrum [387] with a wide water vapor absorption band in this part of the spectrum, this fact and the possible differences between methods used to measure the absolute spectral sensitivity of standard cells could account for the discrepancies between the measured efficiencies of Soviet and Hungarian solar cells.

The differences recorded when simulators are adjusted against different standard cells indicate that it is essential to use a unified standard spectrum of terrestrial solar radiation in the calibration of standard cells. The recently suggested standard terrestrial spectrum (AM1.5 conditions), which has been agreed internationally [391,393], seems to be the only correct solution to the complex problem of calibration of terrestrial solar cells because it will enable us to compare the efficiency and quality of solar cells and batteries manufactured in different countries and by different companies [363,392,393,420,423].

The Makarova–Kharitonov AM0 spectrum [358] can be used not only to calibrate standard cells for measurements of the efficiency of space solar cells and batteries, but also to adjust laboratory

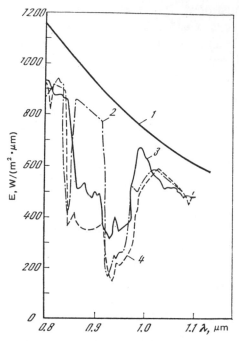

FIG. 4.20. Energy distribution in the region of one of the principal absorption bands of water vapor in the atmosphere (0.8-1.1 μm) under different conditions: 1, 2) standard extra-atmospheric and terrestrial AM1.5, respectively; 3) AM1; 4) AM2.

and industrial Sun simulators and to predict the performance of solar batteries outside the atmosphere. It is much more difficult to take into account the wavelength, flux, and time-dependent component of solar radiation that is due to reflection by clouds and the ground and is efficiently converted by two-sided and infrared-transparent solar cells [5,110,149]. However, the calculated and measured results show that it will not be long before we can accurately predict the possible increase in the current produced by solar cells on low-orbit satellites due to this component of extra-atmospheric solar radiation [143,354].

We have already frequently emphasized that a solar cell used as a standard must have similar spectral, photoelectric, and optical characteristics to those of the solar cells to be examined. For example, when the efficiency of a batch of n^+-p-p^+ silicon solar cells with a shallow p–n junction is to be investigated, the standard cell must be selected from among them, whereas, for solar cells made from the newer semiconducting materials, the standard cell must be produced from the same semiconducting material and with the same layer thickness as in the structure of the solar cell with which it is to be compared.

However, the example given in Section 4.4 of measurements performed from the Venera stations shows that there is another approach, namely, it is possible to develop a stable solar cell, for example, a silicon cell with a relatively deep p–n junction, and include in its certified data the conversion coefficients which must be used when the cell is used as a standard to adjust the Sun simulator for measurement of the parameters of solar cells made from a different semiconducting material or from the same material but of different construction.

It was established during the Venera-13 and Venera-14 missions that, when the extra-atmospheric solar-cell parameters (AM0, 1360 W/m^2) were determined under a Sun simulator incorporating a hot-filament lamp without spectrum correction and producing 1000 W/m^2, the conversion factor for a silicon standard cell with a deep p–n junction (1.0-1.2 μm) was 1.0, whereas the corresponding figure for a silicon cell with a shallow p–n junction (0.3-0.5 μm) was 1.12-1.13. For AlGaAs–GaAs solar cells, the conversion factor was 1.11-1.12, 1.2-1.21, and 1.54 for AlGaAs layer thickness of 15, 10, and 1 μm, respectively.

We can proceed in a similar way in the case of the calibration of Sun simulators used to measure the parameters of terrestrial solar cells. The calibration certificate of the standard cell must then include the value of the coefficient used to convert from AM0 to standard terrestrial conditions (AM1.5), which is either measured or calculated as in the previous case for the same working temperature, with the short-circuit current for AM0 and AM1.5 reduced to the corresponding radiation flux density. The values of such conversion coefficients can readily be deduced from high-altitude measurements, or from the absolute spectral sensitivity of the given standard cell and the standard AM0 and AM1.5 spectra. For silicon solar cells with shallow p–n junctions (0.3-0.5 μm), the conversion coefficient between AM0 and the standard terrestrial conditions corresponding to AM1.5 has been shown by measurement and calculation to be 1.13-1.14. For the AlGaAs—GaAs solar cells, it amounts to 1.26, 1.24, and 1.18 when AlGaAs layer thickness is 15, 10, and less than 1.0 μm, respectively. For thin-film solar cells with the Cu_2S—CdS heterostructure and spectral sensitivity as shown in Fig. 2.23 (curve 2), the AM0 → AM1.5 conversion coefficient is 1.04, whereas for cells with the ITO—Si heterostructure (curve 1, Fig. 2.29), the conversion coefficient is shown by calculation to be 1.10-1.11.

The results reported in [441] are similar qualitatively (and, in some cases, quantitatively, too). The spectral sensitivity of silicon solar cells was measured with a filter monochromator, using a powerful hot-filament tungsten lamp, and was then converted to the solar emission spectra under AM0 [369] and AM1 [376] conditions. The integrated short-circuit current of the solar cells and the AM0 → → AM1 conversion factors were determined from these data. The latter were found to be 1.08 for the ordinary cells without coatings, 1.14 for cells with a texturized nonreflecting surface produced by selective chemical etching, 1.15 for ordinary cells with a silicon dioxide antireflective coating, 1.16 for the same cells with a deeper p—n junction designed for terrestrial applications, and 1.18 for cells with a shallow p—n junction and an antireflective coating of tantalum pentoxide. All the cells were made from single-crystal silicon and the conversion factors quoted above were obtained as averages over batches of two, four, five, nine and eleven cells, respectively. The calculated values were verified by measurements in natural sunlight (air mass m = 1.7; 850-950 W/m^2), and the measured short-circuit currents were reduced to AM1 conditions (about 1000 W/m^2).

Field measurements of the parameters of solar batteries under terrestrial conditions

The parameters of cells and batteries under natural sunlight can be determined in both direct and total flux. In the former case, standard solar cells equipped with collimating tubes (angular field of view not more than 3°) are lined up with the Sun and then replaced by the cells to be examined. The direct solar flux density is additionally determined with a pyrheliometer. The following conditions must be observed during measurements: the flux density on

the cells must be in the range 750-900 W/m^2, the air mass must be
between 1 and 2, the sky must be clear and blue, indicating the
absence of aerosol scattering, and the product of the air mass and
turbidity coefficient must be not more than 0.25 [392].

Measurements in the total flux [389-392] are performed with
horizontally placed cells. The radiation flux density is additionally
determined by a pyranometer. At the same time, if the necessary
equipment is available, the wavelength dependence of the total
flux density is also investigated. The calibration value corresponding
to standard conditions is calculated from the measured and standard
values of radiation flux density, given in the calibration certificate
of the standard cell. When the spectral distribution of the total
flux at the time of measurement is known and deviates substantially
from the standard value (AM1.5 conditions), the results of measurements must be suitably corrected.

The following conditions must be observed: clear weather with
direct solar flux density not less than 800 W/m^2, total scattered
flux not more than a quarter of the direct flux, and altitude of the
Sun above the horizon not less than $54°$. In all cases, the voltmeters
used must have internal resistance of not less than 10 $k\Omega/V$. The
voltage drop in the circuit used to measure the short-circuit current
must not be more than 20 mV per cell.

It is desirable to measure the short-circuit current at near-zero
voltage, which can be done by using a backing voltage (from a
separate electronically stabilized unit) to compensate the voltage
drop across the series resistance.

The current-voltage characteristic can be recorded manually
or automatically, with the data plotted on an XY stripchart recorder
or stored in a minicomputer.

It is important to note that terrestrial solar cells spend only
a small proportion of time working under near-standard conditions
(AM1.5 and the corresponding components of the atmosphere).
The altitude of the Sun above the horizon varies in the course of
the day and there is a corresponding variation in the air mass, the
depth of absorption bands due to water vapor, ozone, and oxygen,
and in the influence of aerosol scattering. Changes in weather conditions, such as the sudden appearance of clouds and rain, produce
still greater departure of the flux density and spectrum from the
standard values.

Many researchers have tried to determine, both experimentally and by calculation, the variation in the principal cell parameters (above all, the efficiency) with variations in the components of the atmosphere and solar spectrum. It is clear that the results of such studies depend not only on differences in the solar spectrum, but also on the nature of the spectral sensitivity of the solar cell made from a particular semiconducting material. The values of the conversion factor between the extra-atmospheric (AM0) and standard terrestrial spectrum (AM1.5) for standard solar cells made from different semiconducting materials and having different structures can be used to estimate the change in the cell characteristics when they are exposed to solar spectra that differ from generally accepted or standard spectra.

It is particularly difficult to isolate the effect of some particular solar parameter, say, the air mass m, on the solar cell characteristics because several parameters will always vary under field conditions. This is probably the reason why field measurements in Cleveland, Ohio, in the United States showed that the efficiency of silicon, gallium arsenide, and cadmium sulfide solar cells did not appear to depend on the air mass m in the range between 1.3 and 4.5 (with simultaneous considerable variations in the radiation flux density for given air mass, due to a rapid transition from a clear sky to cloud or mist) [457]. The measurements were performed between December 1974 and March 1975, when the variations in solar radiation penetrating the atmosphere were small and the cell current was practically proportional to the direct flux density estimated from the readings of a pyrheliometer independently of the air mass, which varied within wide limits.

The experimental studies performed at the Comsat laboratories at Clarksburg, Maryland, between November 1975 and the first ten days of January 1976 showed that the efficiency of silicon solar cells was a function of air mass [458]. The solar flux density was determined with a pyranometer which, whenever it was necessary to separate the direct component from the total radiation, was covered by a black disk, 4 cm in diameter, at a distance of 30 cm from the sensitive surface. The experimental data were confirmed by calculations [458], using measurements of the absolute spectral sensitivity of the cells and terrestrial solar spectra for different parameters of the atmosphere [459,460]. The AM0 extra-atmospheric comparison spectrum was taken to be the Johnson spectrum [369]. Silicon solar cells with shallow p—n junctions made by the Comsat Company were used throghout. Similar results were obtained for gallium arsenide cells [458].

Despite the much lower output power recorded under terrestrial conditions (because of the lower flux density) as compared with the power generated by solar batteries outside the atmosphere, the conversion efficiency can be greater in the former case by almost 20%, depending on the nature of the incident spectrum and the spectral sensitivity of particular solar cells [458]. If we use the smoothed envelope of the AM1 spectrum (the selective absorption bands due to water vapor, ozone and oxygen are then ignored), we obtain solar cell efficiencies that are equal to those under AM0 conditions. For the real AM1 spectrum with selective absorption bands, the cell efficiency is higher by 10% than the AM0 efficiency (conversion factor 1.5), whereas the increase is 15% for AM2 (conversion factor 1.15) and 16% for AM3 (conversion factor 1.16).

Measurements have shown that the conversion factor for AM2 conditions and relatively clear weather is 1.11-1.12 (it is the same for both the total and direct radiation), whereas for misty days it is 1.15 to 1.19.

It is thus clear that the terrestrial parameters of solar cells have the following important feature: as the air mass, mistiness, cloudiness, and number of raindrops increase, the efficiency of solar cells usually substantially increases, although the absolute amount of power generated by them decreases [458]. The reason for this can readily be seen by comparing the spectral energy distribution curves for different air masses (see Fig. 4.2). As the air mass increases from 1 to 5, the flux density decreases, but the maximum of the solar radiation transmitted by the atmosphere shifts to the right and comes closer to the spectral sensitivity maximum of silicon and gallium arsenide cells [380].

Of course, this obvious relationship holds for direct solar radiation. The spectrum of scattered radiation depends on the ratio of the diameter of particles in the mist or cloud, or the raindrop diameter, to the wavelength of solar radiation. However, the increase in the conversion factor from 1.11-1.12 under good weather conditions to 1.15-1.19 in poor weather [458] suggests that the ratio of the wavelength of solar radiation to the dimensions of the scattering particles may be such that it is mostly shortwave radiation that is absorbed and retained in the atmosphere, so that the transmitted spectrum comes closer to the spectral sensitivity of solar cells.

Molecular and aerosol scattering under clear skies produces the reverse effect, i.e., a reduction in the conversion factor between AM0 and terrestrial conditions. When diffuse scattering under a clear sky is taken into account, this should bring the terrestrial spectrum closer to the extra-atmospheric spectrum because the spectral distribution of this scattering is concentrated in the short-wave part of the spectrum (see Figs. 4.4 and 4.5). Calculations support this conclusion: for modern silicon solar cells, with high sensitivity at shorter wavelengths, the conversion factor between AM0 and AM1.5 is 1.17 when only the direct flux of solar radiation is taken into account, but this value falls to 1.14 in the case of the total flux (including diffuse scattering under clear sky conditions [420]. The same tendency was reported in [420] for gallium arsenide and cadmium sulfide cells.

In our view, calculations provide a much clearer definition of the effect of the air mass, the individual spectral intervals in terrestrial solar radiation, and the properties of the cells on the value of the conversion factors. Calculations performed by E. S. Makarova on the conversion factors between AM0 and a number of known terrestrial spectra are listed in Table 4.3.*

I. S. Orshanskii has calculated the conversion factor for two types of silicon solar cell with very different spectral sensitivity (due to the difference in the depth of the p—n junction) for given composition of the atmosphere (international AM1.5 spectrum [391] with only one parameter varying, namely, the air mass, and all other parameters held constant as follows: thickness of the precipitated layer of water vapor 2.0 cm, ozone 3.4 mm; $\beta = 0.12$, $\alpha = 1.3$). The results were as follows:

m	1	1.5	2	3	5	10
$l_d \simeq 2.5\ \mu m$	1.155	1.175	1.2	1.23	1.26	1.27
$l_d \simeq 0.7\ \mu m$	1.12	1.135	1.145	1.155	1.155	1.125

*The AM1.5 spectrum used to obtain the data in Table 4.3 is shown in Fig. 4.5 (curve 2) [391], while the diffuse component (molecular and aerosol scattering) is shown in Fig. 4.3 (curve 2) [382,383]. The AM3 and AM5 spectra (direct radiation) are shown in Fig. 4.2 (curves 4 and 6) [380].

TABLE 4.3

Conversion Factors between AM0 and Differential Terrestrial
Solar Spectra for the Leading Types of Solar Cell

Cell	Total spectrum, $\lambda = 0.25-3.0 \mu m$			
	AM1.5 (direct)	AM1.5 (direct + + diffuse)	AM3 (direct)	AM5 (direct)
Si with shallow p—n junction, $l_d \cong 0.3 \mu m$ (Fig. 4.18, curve 6)	1.137	1.113	1.114	1.076
Si with deep p—n junction, $l_d \cong 1.5 \mu m$ (Fig. 4.18, curve 3)	1.186	1.175	1.193	1.180
GaAs—AlGaAs, 10 μm thick (Fig. 4.19, curve 4)	1.242	1.205	1.244	1.246
GaAs—AlGaAs, 1 μm thick (Fig. 4.19, curve 3)	1.185	1.173	1.168	1.148
ITO—SiO$_x$—nSi (Fig. 2.29, curve 1)	1.106	1.101	1.107	1.087
Cu$_2$S—CdS (Fig. 2.23, curve 2)	1.042	1.08	0.097	0.873

Cell	Part of spectrum $\lambda = 0.3-1.2 \mu m$			
	AM1.5 (direct)	AM1.5 (direct + + diffuse)	AM3 (direct)	AM5 (diffuse)
Si with shallow p—n junction, $l_d \cong 0.3 \mu m$ (Fig. 4.18, curve 6)	1.115	1.071	1.145	1.164
Si with deep p—n junction, $l_d \cong 1.5 \mu m$ (Fig. 4.18, curve 3)	1.163	1.130	1.226	1.276
GaAs—AlGaAs, 10 μm thick (Fig. 4.19, curve 4)	1.218	1.159	1.270	1.347
GaAs—AlGaAs, 1 μm thick (Fig. 4.19, curve 3)	1.162	1.129	1.195	1.241
ITO-SiO$_x$—nSi (Fig. 2.29, curve 1)	1.085	1.059	1.139	1.175
Cu$_2$S—CdS (Fig. 2.23, curve 2)	1.022	1.039	0.99	0.944

Uncertainty in the measured parameters of solar cells
and batteries under simulated and field conditions

Even the highest-quality simulators do not reproduce with perfect precision the optical parameters of the standard solar radiation. Uncertainties in the measured electrical parameters of solar cells and batteries due to this factor can be substantially reduced by using standard cells to adjust Sun simulators. This uncertainty is *reduced* rather than removed altogether because the optical characteristics of solar cells under examination may differ from those of the standard cell. In 1976, the European Space Center and the Comsat Company collaborated in examining 15 solar cells manufactured by Comsat under 9 simulators from different laboratories and organizations in the USA and Western Europe. The spread in the values of the short-circuit current, obtained on different simulators, was 13-15% [461]. This indicates the importance of estimating the precision of measurements made on particular types of solar cell under particular simulators.

The measurement uncertainty consists of spectral, angular, surface, temperature, and temporal components [462]. It has been found that the uncertainty depends equally on the difference between the simulator spectrum and the true solar spectrum and the difference between the spectral sensitivities of the cell under investigation and the standard cell.

In simulators with highly nonuniform illumination, the uncertainty can be reduced by introducing a correction for the difference between the flux density on standard and working cells (or groups of cells). The correction is calculated from a detailed flux density distribution chart for the working field and the true ratio of the reference cell area (for one or a group of cells) and the area of the object under investigation (cell, group of cells, or large battery). The correction is equal to the ratio of the average flux density over the area of the object under investigation to the area of the standard receiver. When the flux density distribution is measured, the area of the standard cell must be at least a quarter of the area of the measured object. A cell of 5 × 5 mm is used in simulators designed for measurements on individual cells. The flux density distribution on solar batteries is determined with a standard group of solar cells of 70 × 75 mm [419]. When the correction is introduced, the

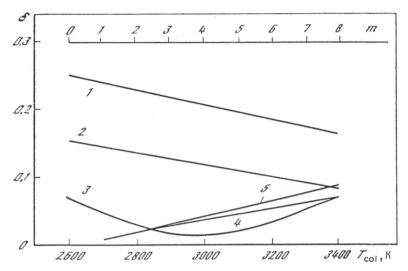

FIG. 4.21. Uncertainty in measurements made under simulators incorporating hot-filament lamps with color temperature T_{col} and under field conditions in terrestrial solar spectra for different air masses, due to the difference between the source spectra and the extra-atmospheric solar spectrum: 1) hot-filament lamp without spectrum correction; 2) hot-filament lamp with water filter, 4 cm thick; 3) S-1 simulator; 4) Sun at altitude of 2000 m above sea level (Mount Aragats in Armenia);   5) Sun at sea level, thickness of precipitated water layer, 2.0 cm [380].

uncertainty does not exceed 2%, even when the nonuniformity of illumination is up to ±10% (in the case of large batteries).

The simplest method of monitoring the spectrum is to perform a periodic check on the blue-red ratio [419]. This can be done by taking alternate measurements with the standard cell, using filters isolating blue and near-infrared radiation [400,461,463] or a dichroic mirror [463].

The uncertainty in the measured short-circuit current of a specific solar cell or battery due to the difference between the spectral distribution of the simulator (which can be the terrestrial solar radiation under nonstandard conditions) and of the Sun (standard extra-atmospheric or terrestrial) can be calculated from the following formula [366,396]:

$$\delta = \frac{I_{sim} - I_s}{I_s} = \frac{\int\limits_0^\infty E_{s\lambda} S_{st}\, d\lambda \int\limits_0^\infty E_{sim\lambda} S_{sc}\, d\lambda}{\int\limits_0^\infty E_{s\lambda} S_{sc}\, d\lambda \int\limits_0^\infty E_{sim\lambda} S_{st}\, d\lambda} - 1 \,,$$

where I_{sim}, I_s is the short-circuit current of the cell under investigation under the simulator and in solar radiation with a standard spectrum, respectively, $E_{s\lambda}$ is the spectral density of the standard solar radiation, $E_{sim\lambda}$ is the spectral flux density produced by the simulator, and S_{st}, S_{sc} is the relative spectral sensitivity of the standard and tested solar cells.

Figure 4.21 shows the uncertainty in the measured current of a solar cell due to the difference between the spectral distribution of the radiation produced by simulators and the standard extra-atmospheric solar spectrum as a function of the color temperature T_{col} of the hot-filament lamp, together with the uncertainties incurred in measurements in terrestrial solar radiation for different values of the air mass m [396].

The spectral uncertainty is small only in the case of simulators with good enough spectrum correction, for example, the S-1 simulator (curve 3), but, in natural sunlight, especially at high altitudes, it is also possible to perform qualitative measurements of solar cell and battery parameters, especially for low values of the air mass (curves 4 and 5).

4.6 Estimating the quality of solar cells and batteries without using Sun simulators

Batteries may become damaged during transport or storage, and it is desirable to have some means of estimating this damage rapidly although not necessarily very accurately. This can be done by using flat panels of light-emitting diodes [464]. Such light diodes have already found applications in metrology, for example, in equipment for measuring low flux densities (GOST 8.273-78). The working state of batteries can readily be estimated even under low illumination by placing such light-emitting diode panels of flat radiators between solar cell panels.

Defects on the outer surface of glass, detachment of coatings, or changes in their interference color produced by unfavorable climatic factors (for example, increased humidity) can also be detected without using Sun simulators or panels for measurements in natural terrestrial sunlight. Changes in the optical properties of batteries due to these defects can be estimated by measuring the reflection coefficient at the point of damage, using a small portable spectrophotometer. The spectral reflection coefficient in the range 0.3-2.5 μm can then be readily converted into the integrated solar absorption coefficient using the appropriate nomograms [46,23]. Because of the increased interest in the determination of the solar absorption coefficient and terrestrial solar spectrum (see Chapter 3), G. A. Gukhman has constructed nomograms (similar to those for the AM0 spectrum [369,356]) which can easily be used to determine α_S for the extra-atmospheric spectral distribution proposed by Makarova and Kharitonov [358,361] and the terrestrial solar spectra under standard conditions [380,382,383,390,391] (Fig. 4.22).

One other method has been used to monitor the quality of solar batteries without using Sun simulators [21]. A dc voltage from an external source is applied to the battery in the forward direction, and the current flowing through the unilluminated battery is measured. It is desirable for this current to be not less than 50% of the short-circuit current produced by the illuminated battery. The forward dark current depends, to a considerable extent, on the temperature of the battery, which is held strictly constant during these measurements. It is also useful to connect the battery to a pulsed source of voltage (pulse separation 10 sec). This method of checking the electrical parameters of batteries has been successfully used in monitoring individual blocks and modules of solar batteries on the Skylab station. It is a good way of estimating the number of parallel-connected groups and modules in a battery, and of searching for any breaks in electrical connections between them. Since the overall series resistance of the unilluminated battery is also a function of the number of series-connected cells, this method can also be used to check the continuity of the series circuit of the cells.

It is important to note that, by monitoring solar battery parameters (for example, after lifetime tests on a Sun simulator) in terms of the measured load current-voltage characteristic, we determine only the overall change in the output power. Laborious checks on the quality of battery components are necessary to identify which particular groups, modules or individual cells, have become defective.

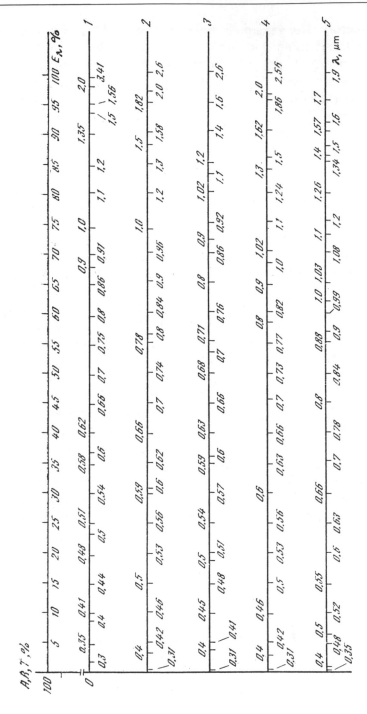

FIG. 4.22. Nomogram for calculating the integrated absorption coefficient α_S of the surface of solar cells and photothermal collectors in extra-atmospheric and terrestrial solar radiation for different values of the air mass: 1) AM0 [358]; 2) AM1.5; 3) AM1.5 + diffuse scattering; 4) AM2 [389]; 5) AM5 [380].

It is much simpler and faster to locate damage such as cracks appearing after thermal cycling by measuring the integrated thermal emission coeffecient of the surface with a scanning infrared thermoradiometer.   Solar cell modules or groups are then arranged to lie in the same plane, and a dc voltage from an external source is applied in the reverse direction. The Joule heat release is greater at points of higher resistance and in regions of peeling, cracking, and breaking, and these can be easily identified by the thermoradiometer [103, 105,449,450]. However, the thermoradiometer cannot be used for solar batteries because the thermal emission from the defective points cannot be distinguished from that of other areas.

4.7 Space applications of standard cells

Standard solar cells made from different semiconducting materials are widely used to adjust Sun simulators for measurements of the parameters of solar cells and batteries under laboratory and manufacturing conditions, and in determinations of the characteristics of terrestrial solar batteries under field conditions. Less frequent but just as important from the scientific point of view is the use of standard solar cells in the study of the characteristics of solar radiation itself, for example, from spacecraft, where the relatively low relative mass, reliability, stability, heat resistance, and small size of semiconductor solar cells make them very convenient and sensitive solar radiation sensors [465].

In the external containers carried by Cosmos-1122 satellites with biological objects on board, solar cell sensors were used to measure the solar flux density on the surface of these objects in relatively low and almost circular orbits around the Earth (mean orbit height 300 km, inclination of orbit 62°), so that the exposure of the biological objects in open space could be monitored. The total amount of time during which cosmic-ray detectors located in the same containers were pointing toward the Sun was also determined.

The satellite was launched into an Earth orbit on September 25, 1979, and the external containers were returned to Earth after about 20 days. The satellite was not stabilized during its mission but executed a slow rotation about an axis whose direction was not controlled. The containers were good enough to protect the objects inside them during descent through the Earth's atmosphere.

The flux density was measured with an independent DKO-2 integrating sensor which was not, however, provided with a telemetry channel. The sensor consisted of two standard silicon solar cells arranged in opposition and connected through a divider to a mercury coulometer and compensating semiconductor diodes [465]. The cells can be alternately exposed by removing a shutter, and the electric current produced by the open solar cell (with the other covered) is passed through the coulometer in the form of a glass capillary with two mercury electrodes separated by a drop of electrolyte, called the coulometer index. When the solar cell is illuminated, a current flows and produces a change in the position of the electrolyte drop within the capillary. This displacement is a measure of the amount of electricity passed through the coulometer, which in turn is proportional to the amount of light incident on the surface of the solar cell [466].

Before it was mounted on the satellite, the DKO-2 sensor and its standard solar cells were calibrated against a Sun simulator. The maximum integrated flux density in space amounted to 0.95 of the total scale length of the coulometer.

Metal plates with selective surface coatings were also placed on the inner lid of the container carrying the scientific equipment of the Cosmos-1129 satellite. Some, for example, white acrylic enamels, were found to darken when exposed to solar radiation. The rate of increase of the integrated solar absorption coefficient of these surfaces has been extensively investigated in space on satellites in the Venera series, using probes aligned with the Sun with sufficient precision [23]. The light flux incident on the coatings and the integrated flux density were additionally estimated from the change in the solar absorption coefficient of these coatings throughout the mission.

If we know the solar absorption coefficient before and after the space mission, we can use its dependence on the flux density or time in solar days (it was assumed that the solar constant was 1360 W/m² [23]) to estimate the interval of time during which the objects inside the containers were illuminated by solar radiation incident at right angles to the surface of the coated plates, and perform a comparison of the two methods of measuring the flux density during the mission.

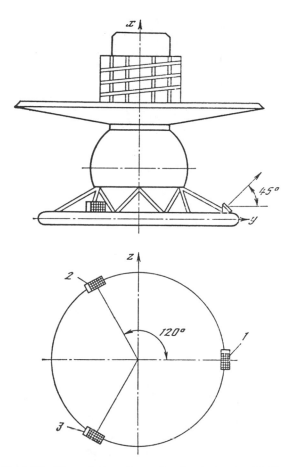

FIG. 4.23. Disposition of small solar batteries (1-3) on the exterior of the Venera-11 interplanetary station.

The Cosmos-1129 measurements of the integrated solar flux incident on the objects in the containers gave 4.6 equivalent solar days as measured by the darkening of the coatings and 4.5 days according to the DKO-2 sensors with standard solar cells.

We note that the data obtained correspond to a solar flux incident along the normal to the surface of the sensors and coated plates. We are therefore dealing with radiation flux equivalent to the direct flux (along the normal), since the *total* true flux is incident on flat surfaces at different angles within the solid angle of 2π.

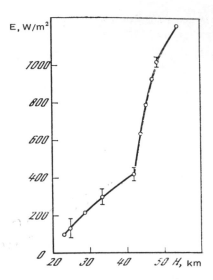

FIG. 4.24. Solar flux density as a function of the height of the descending interplanetary station above the surface of Venus, obtained with calibrated silicon solar batteries.

These investigations have, therefore, established with sufficient precision that the containers carried by the unstabilized Cosmos-1129 satellite (about 20 days in orbit) were exposed to solar radiation equivalent to the direct solar flux falling along the normal to the surface for 4.5-4.6 days.

Semiconductor solar cells that convert solar radiation directly into electric power can be used, as already mentioned, not only as the sources of power for spacecraft, but also as sensors of direct and reflected light, when the Earth is investigated from space [149,354], and in studying the transparency of the atmospheres of other planets, for example, Venus. Most of the spacecraft entering the Venusian atmosphere were therefore equipped with small solar batteries calibrated on Earth under laboratory conditions [467]. The information provided by these batteries not only complemented the spectral measurements of the transparency of the Venusian atmosphere [468], but was also used to estimate the diffusivity of the light flux in that atmosphere. During the descent of Venera-5 on 16 May 1969, bursts of illumination were recorded for the first time on the night side of the planet [469], which together with light measurements performed by Venera-9 and Venera-10, and measurements of the radio noise due to electrical discharges on Venera-11 and Venera-12, led to the conclusion that there were lightning discharges in the atmosphere of this planet [470].

Small batteries have been developed to withstand a load of 300g (g is the gravitational acceleration) as they enter the atmosphere of the planet, and also high temperatures and pressures (up to 300°C and 10^7 Pa). As a rule, each battery consists of 28-30 individual calibrated silicon solar cells connected in series with one another and with a load resistor producing an output voltage of 0-6 V when the solar flux density varies from 50 to 1500 W/m².

The small solar battery and the other equipment on the Venera station were cooled down to about −15°C [469] prior to descent, so that an acceptable temperature could be established for the solar cells on most of the flight paths. A temperature correction was therefore introduced in the analysis of the results.

On 25 December 1978, the descending Venera-11 carried three small solar batteries mounted on its exterior (Fig. 4.23). They were used to measure the solar flux density in the atmosphere of Venus, and showed that the flux density in the Venusian atmosphere was quite high (Fig. 4.24). This agrees with [468,471] and suggests that solar batteries will be capable of operating efficiently as power sources on, for example, balloon-borne stations floating in the Venusian atmosphere [467].

The uncertainties indicated by the vertical bars in Fig. 4.24 are very small, so that the flux is almost completely diffuse in character, and there is no direct component of solar radiation at these altitudes in the Venusian atmosphere. Similar experiments were performed, and analogous results to those shown in Fig. 4.24 were recorded, during the descent of Venera-13 and Venera-14 in March 1982.

Conclusion

No one now doubts the great scientific and applied value of photovoltaic power engineering and the associated optics and metrology [1-5,472]. Extensive work is being done on the development of new solar cells (for example those using amorphous germanium-silicon [473] and silicon-selenium alloys [474]) and selective coatings. Although some of the recent work [475-477] has simply repeated research done 15 years ago [289-296,309], several publications have described methods of deposition of improved coatings [478, 479] which should be widely used in the near future. There has been promising work on evacuated flat-plate collectors [480], lighter solar batteries for space applications, higher radiation stability, and stability to laser radiation [481,482]. All these developments are finding extensive application in the technology and metrology of solar cells and batteries [483-486]. New Sun simulators [487] are being developed and the value of the solar constant is becoming increasingly more accurate [488]. There have been considerable advances in antireflective coatings with depth-dependent refractive index [489], and thin-film cascade cells made from amorphous and polycrystalline silicon with an efficiency of 12.1% have been produced [490]. Laboratories for solar cell metrology are being organized [491]. The international AM1.5 solar spectrum with strictly defined atmospheric parameters is beginning to be part of everyday measurement practice [492, 493]. The spectral composition and flux density of the standard terrestrial solar spectrum, which must be used in measurements of the efficiency of solar cells and batteries, is being measured with increasing accuracy.

The Solar Energy Institute in Colorado, USA, has suggested an AM1.5 standard spectrum of direct solar radiation [494], which is slightly different from that adopted at present [391,393], and a spectrum of total AM1.5 solar radiation incident on a surface at 37° to the horizontal [494]. This spectrum is so enriched with short-wave radiation due to the diffuse component and solar radiation reflected from the Earth (the Earth's albedo is assumed to be 0.2) that its spectral distribution, as indicated by the blue-red ratio, is very close to AM0.

Terrestrial photogenerators exploit the total rather than the direct solar radiation, but the use of the total radiation in efficiency measurement is made difficult by factors such as, for example, the dependence of the cell reflection coefficient and photocurrent on the angle of incidence of solar rays on the cell surface. When the total radiation spectrum is taken as standard, the standard and tested cells must have not only practically identical spectral sensitivity, but also comparable angular dependence of the spectral reflection coefficient and photocurrent.

Optical and metrological studies of the properties of solar cells constitute a new topic in energy conversion that is important from both theoretical and practical points of view.

Without the development of the optics and metrology of solar cells it would have been impossible to measure accurately the efficiency of terrestrial and space solar cells and batteries, or to optimize their characteristics for different working conditions.

As already noted, the UN International Electrotechnical Commission decided in 1982 to define the standard spectrum in terms of solar cell measurements in the direct flux under AM1.5. The efficiency of solar cells made from different semiconducting materials, and operating under different climatic and geographic conditions, can then be estimated using this spectrum of solar radiation on the Earth, in the direct flux density of 834.6 W/m^2, together with other theoretical and experimental spectra of direct solar radiation. However, it is still necessary to remember that solar cells working without concentrators transform both direct and diffuse solar fluxes, including the component due to molecular Rayleigh scattering by the atmosphere. The diffuse component may be very considerable even on clear days.

These factors led Technical Committee No. 82 of the UN Electrotechnical Commission to recommend the following standard: measurements on solar cells for terrestrial applications should be made in a total flux of direct terrestrial solar radiation of 1000 W/cm^2, the spectrum being typical for AM1.5 conditions and albedo of underlying surface equal to 0.2. The actual spectral distribution of this standard terrestrial solar spectrum, which includes both the direct and the diffuse components, is given in tabular form in Appendix 3 at the end of this book.

The recommended spectrum is increasingly used in determinations of the efficiency of terrestrial solar cells and batteries. This is clear, for example, from reports published in the Proceedings of the 18th IEEE Photovaltaic Specialists' Conference [495].

Many unforseen difficulties will have to be overcome in the course of development, improvement, and testing of different types of solar cell.

For example, it has recently become clear that atomic oxygen present in space around the Earth is capable of actively damaging the polymer film on which solar cells are mounted on most American spacecraft, whereas electrical discharges between the dielectric coatings on the front and rear of solar cells (which occur because these surfaces become charged up and a potential difference is set up between them) can produce malfunction in some batteries.

Methods for solving these problems have already been outlined. It is probably best to replace the polymer supporting panels with glass cloth. Surface charges can be prevented from accumulating on the dielectric coatings by introducing components into the polymers or glasses that increase their bulk conductivity, and by coating the inner and other surfaces with a transparent conducting film of indium and/or tin oxides (these films must, of course, be in electrical contact with one another and with the body of the apparatus).

Transparent conducting films of indium and tin oxides are widegap semiconductors that are very suitable as photoactive optical windows for solar cells based on heterostructures. They are being increasingly used in new silicon, indium phosphide, and amorphous silicon solar cells. The efficiency of solar cells based on heterostructures consisting of a film of a mixture of tin and indium oxides plus indium phosphide is already 16% in excess. The cells have high radiation stability and are relatively simple to fabricate.

Recent conferences attended by Soviet specialists (meeting of SEC countries, Ashkhabad, September 1986) and by American workers (17th and 18th IEEE conferences on photovoltaic solar energy conversion in the USA; published in 1985—1987) have demonstrated that considerable theoretical and experimental advances have already been made in this new and rapidly developing branch of science and technology.

In particular, proposals have been made for solar cells consisting of superlattices formed by exceedingly thin alternate epitaxial films of gallium arsenide and solid solutions of aluminum, gallium, and arsenic; gallium, indium, and arsenic; and gallium, antimony, and arsenic. Apart from high efficiency, these cells have exceptionally high radiation stability because of the virtually complete collection of minority carriers when the carrier diffusion length after exposure to radiation is only 300-500 Å.

Inexpensive cells have been made from silicon strip and from silicon produced by the casting method; they incorporate a diffused junction and a back barrier, produced by baking-on printing paste and doping from suitable solutions. The efficiency of such cells has reached 16-17%.

Many different cascade cells has been developed, including thin-film multilayer structures based on amorphous silicon. It is hoped that efficiencies of 18% will be reached by 1988.

In practice, an efficiency of 11% has been achieved for copper and cadmium chalcogenides. These cells are based on the $CdZnS$-$CuInSe_2$ heterostructure, which means that carriers can be efficiently collected by exploiting the depth-dependent gap width in the first of these materials.

The most striking development in recent years is undoubtedly the appearance of new solar cells designs using single crystal silicon — the veteran of solar energy conversion. The efficiency of these new cells (measured under AM1.5, with solar flux of 1000 W/m^2, and concentration ratios of up to 500) has exceeded 19% and has reached 22.4 for one of these structures!

Three cell designs based on single crystal silicon, and developed in Australia and the USA [496—498], are of particular interest.

The first of these is a solar cell with point contacts on the rear surface. The base plate, 70-100 μm thick, is made from silicon grown by the crucible-free zone refining method. It has a resistivity of 200 $\Omega \cdot$cm and long carrier lifetime (about 1 msec). The light-receiving surface of the cell does not have the usual diffused p—n junction. It can be polished or textured, and is covered by a passivating silicon oxide film, 1100 Å thick, which sharply reduces the rate of surface recombination of carriers. On the rear surface there are minute (10×10 μm) highly doped p- and n-type spots, only 30 μm apart. The total number of these microspots on an area of 0.8×0.8

cm is 73441! Each microspot has an aluminum contact of 5×5 μm (the junction between the highly doped silicon and aluminum is made by a palladium—silicon alloy). The aluminum contacts with the n- and p-type microsopots are insulated from one another by aluminum oxide layers and the current leads on the rear surface resemble two interlocking combes separated by a small gap. It was this particular design that was found to have an efficiency of 22.4% for a concentration ratio of 150. It is expected that this figure will rise to 27—28% for a concentration ratio of 500.

These cells with point contacts on the rear surface have high open-circuit voltages U_{oc} (0.81 V for a concentration ratio of 150). The reason for these high values of U_{oc} in the case of silicon is the low level of recombination losses of all types: bulk recombination is reduced by using a base with long carrier lifetime, and surface recombination is reduced by the passivating oxide film on both front and rear surfaces; recombination on the contacts is reduced by reducing the contact area. The current generated by this type of cell is also quite high. This is facilitated by the fact that the front surface is not obstructed by contact strips and all the incident light is captured by the structure (aluminum layers above the passivating layers on the rear surface act as good light reflectors, so that multiple reflections are produced between the upper textured surface and the rear reflecting surface).

The second design is the metal—dielectric—semiconductor structure with a p—n junction. It usually includes a low-resistance base layer of 0.1-0.2 $\Omega \cdot$cm, grown by the crucible-free zone refining method that produces a high degree of structural perfection so that the diffusion length of minority carriers is quite large (about 200 μm). The very thin passivating film deposited on both the upper surface of the cell, which intercepts the incident light, and under the contacts on this surface (where its thickness is only 30 μm) produces a substantial reduction in surface recombination in the near-contact regions because carrier transport occurs by electron tunneling from metals such as Ti, Al, and Mg whose work function is lower than that of silicon.

The solar cell with passivated emitter region has the same properties, namely, high open-circuit voltage and low recombination losses. The passivating film in this design is also very thin. Nevertheless, it is capble of suppressing recombination on the surface and under the contacts which, in this design, occupy a very small fraction of the illuminated area. At the same time, the small thickness of the SiO_2 film does not affect the antireflective properties of the TiO_2, SiO_2 or ZnS, MgF_2 coatings. Calculations performed for such structures have shown that a two-layer coating in the form of 475 Å of TiO_2 and 986 Å of SiO_2, deposited on top of the SiO_2 passivating film 60-100 Å thick will produce a 50.4% increase in the short circuit current of solar cells. Ti, Pd, and Ag are used for contacts in the last two designs.

Improvements in the design and technology of solar cells that have been achieved recently include the production of the "velvet" and other surface microstructures by a scanning laser beam, and also metal contact layers produced by laser decomposition of multicomponent printing pastes, and rigidly attached to the semiconductor (the paste is deposited in advance on the front and back surfaces of the semiconductor).

Solar power engineering has a bright future. There is no doubt that this method of solar energy conversion will be increasingly important in the life of humanity.

At the exhibition of scientific and technological achievements of the SEC countries in the field of solar power engineering, which accompanied the Ashkhabad conference of 1986, our colleagues from Turkmenia have demonstrated a field solar electrolyzer. The current generated by a portable solar battery released hydrogen gas from the water collected from a running brook, and a blue flame was seen to burn with some vigor on the gas plate next to the system. Solar and hydrogen power are ecologically clean, convenient, noiseless, and inexhaustible. They were seen to work agreeably together, just as they surely will work for the benefit of humanity, but on much larger scale in the 21st century.

Appendix

1

Extra-atmospheric Solar Spectrum [358]

λ, μm	E_λ, W/(m$^2 \cdot \mu$m)	λ, μm	E_λ, W/(m$^2 \cdot \mu$m)	λ, μm	E_λ, W/(m$^2 \cdot \mu$m)	λ, μm	E_λ, W/(m$^2 \cdot \mu$m)
0.295	567	0.410	1740	0.550	1870	1.00	735
0.300	530	0.415	1770	0.560	1820	1.05	658
0.305	589	0.420	1760	0.570	1820	1.10	595
0.310	668	0.425	1710	0.580	1820	1.15	538
0.315	732	0.430	1650	0.590	1790	1.20	487
0.320	782	0.435	1690	0.600	1750	1.25	441
0.325	898	0.440	1830	0.610	1730	1.30	393
0.330	972	0.445	1940	0.620	1690	1.35	366
0.335	959	0.450	2030	0.630	1660	1.40	337
0.340	959	0.455	2060	0.640	1630	1.45	312
0.345	972	0.460	2070	0.650	1590	1.50	287
0.350	1006	0.465	2060	0.660	1560	1.55	267
0.355	1006	0.470	2050	0.670	1550	1.60	244
0.360	1000	0.475	2080	0.680	1520	1.65	223
0.3625	1040	0.480	2070	0.690	1500	1.70	204
0.3675	1070	0.485	1990	0.700	1480	1.80	168
0.370	1080	0.490	1970	0.720	1410	1.90	135
0.375	1060	0.495	2010	0.740	1330	2.0	111.5
0.380	1060	0.500	2000	0.760	1280	2.2	78.5
0.385	1020	0.505	1990	0.780	1220	2.4	57.3
0.390	1040	0.510	1980	0.800	1160	2.6	43.0
0.395	1130	0.520	1900	0.850	1020	2.8	32.8
0.400	1430	0.530	1930	0.900	897	3.0	25.4
0.405	1670	0.540	1900	0.950	823		

2

AM1.5 Terrestrial Solar Spectrum [391] (E = 834.6 W/m², Thickness of the Layer of Precipitated Water Vapor 2 cm, Ozone 0.34 cm, β = 0.12 α = 1.3)

λ, μm	E_λ, W/(m²·μm)	λ, μm	E_λ, W/(m²·μm)	λ, μm	E_λ, W/(m²·μm)
0.295	0	0.365	355.88	0.435	890.55
0.305	1.32	0.375	386.80	0.445	1077.07
0.315	20.96	0.385	381.78	0.455	1162.43
0.325	113.48	0.395	425.18	0.465	1180.61
0.335	182.23	0.405	751.72	0.475	1212.72
0.345	234.43	0.415	822.45	0.485	1180.43
0.355	286.01	0.425	842.26	0.495	1253.83
0.505	1242.28	0.830	923.87	1.288	345.69
0.515	1211.01	0.835	914.95	1.314	284.24
0.525	1244.87	0.8465	407.11	1.335	175.28
0.535	1299.51	0.860	857.46	1.384	2.42
0.545	1273.47	0.870	843.02	1.432	30.06
0.555	1276.14	0.875	835.10	1.457	67.14
0.565	1277.74	0.8875	817.12	1.472	69.89
0.575	1292.51	0.900	807.83	1.542	240.85
0.585	1284.55	0.9075	793.87	1.572	226.14
0.595	1262.61	0.915	778.97	1.599	220.46
0.605	1261.79	0.925	217.12	1.608	211.76
0.615	1255.43	0.930	163.72	1.626	211.26
0.625	1240.19	0.940	249.12	1.644	201.85
0.635	1243.79	0.950	231.30	1.650	199.68
0.645	1233.96	0.955	255.61	1.676	180.50
0.655	1188.32	0.965	279.69	1.732	161.59
0.665	1228.40	0.975	529.64	1.782	136.65
0.675	1210.08	0.985	496.64	1.862	2.01
0.685	1200.72	1.018	585.03	1.955	39.43
0.695	1181.24	1.082	486.20	2.008	72.58
0.6983	973.53	1.094	448.74	2.014	80.01
0.700	1173.31	1.098	486.72	2.057	72.57
0.710	1152.70	1.101	500.57	2.124	70.29
0.720	1133.83	1.128	100.86	2.156	64.76
0.7277	974.30	1.131	116.87	2.201	68.29
0.730	1110.93	1.137	108.68	2.260	62.52
0.740	1086.44	1.144	155.44	2.320	57.03
0.750	1070.44	1.147	139.19	2.338	53.57
0.7621	733.08	1.178	374.29	2.356	50.01
0.770	1036.01	1.189	383.37	2.388	31.93
0.780	1018.42	1.193	427.85	2.415	28.10
0.790	1003.58	1.222	382.57	2.453	24.96
0.800	988.11	1.236	383.81	2.494	15.18
0.8058	860.28	1.264	323.88	2.537	2.59
0.825	932.74	1.276	344.11		

3

Standard Spectrum of the Total Flux of Solar Radiation on the Earth under
AM1.5 (E_{tot} = 1000 W/m^2, Thickness of Precipitated Water Vapor and
of Ozone 1.42 cm and 0.34 cm, respectively).

λ, μm	E_λ, W/(m$^2 \cdot \mu$m)	λ, μm	E_λ, W/(m$^2 \cdot \mu$m)
0.3050	9.5	0.7575	1175.5
0.3100	42.3	0.7625	643.1
0.3150	107.8	0.7675	1030.7
0.3200	181.0	0.7800	1131.1
0.3250	246.8	0.8000	1081.6
0.3300	395.3	0.8160	849.2
0.3350	390.1	0.8237	785.0
0.3400	435.3	0.8315	916.4
0.3450	438.9	0.8400	959.9
0.3500	483.7	0.8500	978.9
0.3600	520.3	0.8800	933.2
0.3700	666.2	0.9050	748.5
0.3800	712.5	0.9150	667.5
0.3900	720.7	0.9250	690.3
0.4000	1013.1	0.9300	403.6
0.4100	1158.2	0.9370	258.3
0.4200	1184.0	0.9480	313.6
0.4300	1071.9	0.9650	526.8
0.4400	1302.0	0.9800	646.4
0.4500	1526.0	0.9935	746.8
0.4600	1599.6	1.0400	690.5
0.4700	1581.0	1.0700	637.5
0.4800	1628.3	1.1000	412.6
0.4900	1539.2	1.1200	108.9
0.5000	1548.7	1.1300	189.1
0.5100	1586.5	1.1370	132.2
0.5200	1484.9	1.1610	339.0
0.5300	1572.4	1.1800	460.0
0.5400	1550.7	1.2000	423.6
0.5500	1561.5	1.2350	480.5
0.5700	1501.5	1.2900	413.1
0.5900	1395.5	1.3200	250.2
0.6100	1485.3	1.3500	32.5
0.6300	1434.1	1.3950	1.6
0.6500	1419.9	1.4425	55.7
0.6700	1392.3	1.4625	105.1
0.6900	1130.0	1.4770	105.5
0.7100	1316.7	1.4970	182.1
0.7180	1010.3	1.5200	262.6
0.7244	1043.2	1.5390	274.2
0.7400	1211.2	1.5580	275.0
0.7525	1193.9	1.5780	244.6

APPENDIX 3 (continued)

$\lambda, \mu m$	$E_\lambda, W/(m^2 \cdot \mu m)$	$\lambda, \mu m$	$E_\lambda, W/(m^2 \cdot \mu m)$
1.5920	247.4	2.3600	62.0
1.6100	228.7	2.4500	21.2
1.6300	244.5	2.4940	18.5
1.6460	234.8	2.5370	3.2
1.6780	220.5	2.9410	4.4
1.7400	171.5	2.9730	7.6
1.8000	30.7	3.0050	6.5
1.8600	2.0	3.0560	3.2
1.9200	1.2	3.1320	5.4
1.9600	21.2	3.1560	19.4
1.9850	91.1	3.2040	1.3
2.0050	26.8	3.2450	3.2
2.0350	99.5	3.3170	13.1
2.0650	60.4	3.3440	3.2
2.1000	89.1	3.4500	13.3
2.1480	82.2	3.5730	11.9
2.1980	71.5	3.7650	9.8
2.2700	70.2	4.0450	7.5

References

1. N. N. Semenov, "Solar energy at the service of humanity," in: Solar Energy Conversion, N. N. Semenov (Editor), Institute of Chemical Physics, USSR Academy of Sciences, pp. 3-6, Chernogolovka, 1981.
2. N. S. Lidorenko, N. N. Gibadulin, G. P. Pochivalin, S. V. Ryabikov, and D. S. Strebkov, "A study of designs for solar electric power stations for terrestrial needs," in: Solar Energy Utilization. Abstracts of Papers read at the Chernogolovka Conference, 17-19 February 1981, Institute of Chemical Physics, pp. 76-78, Chernogolovka, 1981.
3. N. S. Lidorenko and M. M. Koltun, "Solar energy conversion on satellites. "Zemlya i Vselennaya, no. 2, pp. 27-31, 1979.
4. N. S. Lidorenko and M. M. Koltun, "Solar electricity," Tekh. Nauka, no. 8, pp. 12-14, 1980.
5. N. S. Lidorenko, V. M. Evdokimov, A. K. Zaitseva, M. M. Koltun, S. V. Ryabikov, and D. S. Strebkov, "New solar cells and their optimization," Geliotekhnika*, no. 3, pp. 3-17, 1978.
6. Yu. P. Maslakovets and V. K. Subashiev (Editors), Semiconductor Energy Converters [Russian translation], Izd. Inostr. Lit., 1959.
7. V. S. Vavilov, A. P. Landsman, and V. K. Subashiev, "Solar batteries," in: Artificial Earth Satellites, no. 2(75), pp. 75-80, USSR Academy of Sciences, Moscow, 1968.
8. M. M. Koltun, "Solar batteries," in: The Great Soviet Encyclopedia, 3rd ed., vol. 24, book 1, p. 417, 1976.
8a. M. M. Koltun, "Photoelectric generator," in: The Great Soviet Encyclopedia, 3rd ed., vol. 27, p. 603, 1977.
8b. M. M. Koltun, "The photocell," in: The Great Soviet Encyclopedia, 3rd ed., vol. 27, p. 607, 1977.

*Geliotekhnika has been translated by Allerton Press, Inc. since Volume I (1965) under the English title Applied Solar Energy.

9. A. F. Ioffe, The Physics of Semiconductors, USSR Academy of Sciences, Moscow-Leningrad, 1957 [English translation by Academic Press, 1961].

10. L. S. Stil'bans, The Physics of Semiconductors, Sov. Radio, Moscow, 1967 [English translation by Plenum Press, 1970].

11. V. I. Fistul, Introduction to the Physics of Semiconductors, Vyssh. Shk., 1975.

12. S. M. Ryvkin, Photoelectric Phenomena in Semiconductors, Fizmatgiz, Moscow, 1963.

13. A. M. Vasil'ev and A. P. Landsman, Semiconductor Photo-converters, Sov. Radio, Moscow, 1971.

14. A. A. Frimer and I. T. Taubkin (Editors), Semiconductor Photoconverters and Energy Conversion [Russian translation], Mir, Moscow, 1965.

15. V. K. Subashiev and M. S. Sominskii, "Semiconductor photo-cells," in: Semiconductors in Science and Technology, USSR Academy of Sciences, vol. 2, pp. 115-216, Moscow-Leningrad, 1958.

16. A. Ya. Gliberman and A. K. Zaitseva, Silicon Solar Batteries, Energiya, Moscow, 1961.

17. L. K. Buzanova and A. Ya. Gliberman, Semiconductor Photocells, Energiya, Moscow, 1976.

18. A. Ya. Gliberman, I. I. Kovalev, and G. A. Chetverikova, "Photoconverters in science and technology," in: Electronics and Its Applications, VINITI, vol. 12, pp. 117-162, Moscow, 1980.

19. V. M. Evdokimov, M. B. Kagan, M. M. Koltun, and A. Kh. Cherkasskii, "Direct conversion of thermal and chemical energy into electric power," in: Solar Batteries, VINITI, Moscow, 1977.

20. A. Kh. Cherkasskii and L. L. Silin, "General and theoretical problems in electric power engineering. New current sources," in: Thermoelectric and Photoelectric Generators, vol. 5, VINITI, Moscow, 1972.

21. H. S. Rauschenbach, Solar Array Design Handbook (The Principles and Technology of Photovoltaic Energy Conversion), Van Nostrand Reinhold Co., New York, 1980.

22. L. B. Kreinin and G. M. Grigor'eva, "Studies of cosmic space," in: Solar Batteries Under Bombardment by Cosmic Radiation, vol. 13, VINITI, Moscow, 1979.

23. M. M. Koltun, Selective Optical Surfaces for Solar Energy Converters, Nauka, Moscow, 1979 [English translation by S. Chomet, Allerton Press, 1981].

24. S. Tolansky, Optical Illusion [Russian translation], Mir, Moscow, 1967.

25. T. S. Moss, Optical Properties of Semiconductors [Russian translation], V. S. Vavilov (Editor), Izd. Inostr. Lit., Moscow, 1961.

26. V. K. Subashiev, G. B. Dubrovskii, and A. A. Kukharskii, "Optical constants and free carrier concentration in highly doped semiconducting materials deduced from reflection data," Fiz. Tverd. Tela (Leningrad), vol. 6, pp. 1078-1081, 1964.

27. H. B. Briggs, "Optical effects in bulk silicon and germanium," Phys. Rev., vol. 77, p. 287, 1950.

28. W. C. Dash and R. Newman, "Intrinsic optical absorption in single-crystal germanium and silicon at 77 K and 300 K," Phys. Rev., vol. 99, pp.115-1155, 1955.

29. M. D. Sturge, "Optical absorption of GaAs between 0.6 and 2.75 eV," Phys. Rev., vol. 127, pp. 768-773, 1962.

30. I. E. Davey and T. Pankey, "Structural and optical properties of thin GaAs films," J. Appl. Phys., vol. 35, pp. 2203-2209, 1964.

31. N. F. Kovtonyuk and Yu. A. Kontsevoi, Measurements of the Parameters of Semiconducting Materials, Metallurgiya, Moscow, 1970.

32. R. A. Smith, Semiconductors [Russian translation], Izd. Inostr. Lit., 1962.

33. N. F. Mott and R. W. Henry, Electronic Processes in Ionic Crystals [Russian translation], Izd. Inostr. Lit., Moscow, 1950.

34. W. G. Spitzer and H. Y. Fan, "Determination of optical constants and carrier effective mass of semiconductors," Phys. Rev., vol. 106, pp. 882-890, 1957.

35. L. E. Howarth and J. F. Gilbert, "Determination of free electron effective mass of n-type silicon," J. Appl. Phys., vol. 34, pp. 236-237, 1963.

36. M. B. Kagan, M. M. Koltun, and A. P. Landsman, "A study of reflection by highly doped gallium arsenide in a wide

spectral interval," Zh. Prikl. Spektrosk., vol. 5, pp. 770-773, 1966.

37. E. E. Gardner, W. Kappallo, and C. R. Gordon, "Measurements of diffused semiconductor surface concentrations by infrared plasma reflection," Appl. Phys. Lett., vol. 9, pp. 432-434, 1966.

38. J. C. Irvin, "Resistivity of bulk silicon and of diffused layers in silicon," Bell Syst. Tech. J., vol. 41, pp. 387-394, 1962.

39. A. A. Kukharskii and V. K. Subashiev, "Determination of certain parameters of highly doped semiconductors from the wavelength dependence of the reflection coefficient," Fiz. Tverd. Tela, vol. 8, pp. 753-757, 1966.

40. A. A. Kukharskii, V. K. Subashiev, and M. B. Ushakova, "A layer of degenerate n-type germanium in heterojunctions," Fiz. Tekh. Poluprovodn., vol. 1, pp. 203-205, 1967.

41. H. Y. Fan, "Infrared absorption in semiconductors," Rep. Prog. Phys., vol. 19, p. 107, 1956.

42. T. M. Golovner, M. B. Kagan, and M. M. Koltun, "Determination of free carrier concentration in highly doped gallium arsenide using the wavelength dependence of optical parameters," Elektron. Tekh. Ser. 2, Poluprovodn. Prib., no. 6, pp. 87-95, 1967.

43. B. P. Kozyrev and O. E. Vershinin, "The determination of the infrared spectral diffuse-reflection coefficients of blackened surface," Opt. Spektrosk., vol. 6, pp. 542-549, 1959.

44. M. M. Koltun and L. S. Kridiner, "A study of the temperature dependence of the emissive power of optical coatings," Geliotekhnika [Applied Solar Energy], no. 4, pp. 41-48, 1974.

45. T. N. Krylova and R. S. Sokolova, "Attachment to the IKS-12 spectrophotometer for reflection coefficient measurements," Zh. Opt.-Mekh. Prom-sti, no. 8, pp. 28-32, 1963.

46. L. F. Drummeter and G. Hass, "Solar absorption and thermal emittance of evaporated coatings," in: Physics of Thin Films, G. Hass and R. E. Thun (Editors) [Russian translation], vol. 2, pp. 254-319, Mir, Moscow, 1967.

47. C. E. Jones and A. R. Hilton, "The depth of mechanical damage in gallium arsenide," J. Electrochem. Soc., vol. 112, pp. 908-911, 1965.

48. T. M. Golovner, V. V. Zadde, A. K. Zaitsev, M. M. Koltun,

and A. P. Landsman, "Shallow silicon p—n junctions produced by phosphorus ion implantation," Fiz. Tekh. Poluprovodn., vol. 2, pp. 720-726, 1968.

49. S. P. Malinova and L. L. Odynets, "Oxidation of silicon p—n junctions," Zh. Fiz. Khim., vol. 39, p. 531, 1965.

50. H. Rupprecht and G. H. Schwuttke, "Effects of drift fields and field gradient on the quantum efficiency of photocells," J. Appl. Phys., vol. 37, pp. 2862-2866, 1966.

51. B. McDonald and G. Goetzberger, "Measurement of the depth of diffused layers in silicon by the grooving method," J. Electrochem. Soc., vol. 109, pp. 141-144, 1962.

52. I. F. Gibbons, A. El-Hoshy, K. E. Manchester, and F. L. Vogel, "Implantation profiles for 40-keV phosphorus ions in silicon single crystal substrates," J. Appl. Phys. Lett., vol. 8, pp. 46-48, 1966.

53. P. V. Pavlov, V. A. Uskov, E. I. Zorin, D. I. Tetel'baum, and A. S. Baranova, "Diffusion of boron in silicon from a layer doped by ion implantation," Fiz. Tverd. Tela, vol. 8, pp. 2782-2784, 1966.

54. E. I. Zorin, D. I. Tetel'baum, Yu. S. Popov, and Z. K. Granitsina, "Properties of a surface layer of n-type germanium bombarded by 40-keV nitrogen ions," Fiz. Tverd. Tela, vol. 6, pp. 2017-2021, 1964.

55. C. S. Fuller and J. A. Ditzenberger, "Diffusion of donor and acceptor elements in silicon," J. Appl. Phys., vol. 27, pp. 544-553, 1956.

56. B. I. Boltaks,   Diffusion in Semiconductors, Fizmatgiz, Moscow, 1961.

57. K. Sato, Y. Ishikawa, and K. Sugawara, "Infrared interference spectra observed in silicon epitaxial wafers," Solid State Electron., vol. 2, pp. 771-781, 1966.

58. V. I. Fistul, Highly Doped Semiconductors, Nauka, Moscow, 1967.

59. Yu. A. Kontsevoi, M. M. Koltun, and A. I. Tatarenkov, Soviet Patent No. 258463, Instrument for Controlling the Quality of Polished Semiconductor Wafers.

60. M. L. Timmons, I. A. Hutchby, S. M. Bedair, and M. Simmons, "The development of high efficiency cascade solar cells:

an overview," in: Proc. Sixteenth Intersoc. Energy Convers. Eng. Conf., Atlanta (GA), 9-14 August 1981, vol. 2, pp. 1642-1644, ASME, New York, 1981.

61. V. S. Vavilov and K. I. Britsyn, "Photoionization quantum yield of silicon," Zh. Eksp. Teor. Fiz., vol. 34, pp. 1354-1355, 1958.

62. V. S. Vavilov, Effect of Radiation on Semiconductors, Fizmatgiz, Moscow, 1963.

63. M. Wolf, "Limitations and possibilities for improvement of photovoltaic solar energy converters. Pt. I. Considerations for Earth's surface operation," Proc. IRE, vol. 48, pp. 1246-1263, 1960.

64. N. S. Lidorenko and D. S. Strebkov, "The anomalous photoelectric effect," Dokl. Akad. Nauk SSSR, vol. 2, pp. 325-329, 1974.

65. B. L. Sater, H. W. Brandhorst, T. I. Riley, and R. E. Hart, "The multiple junction edge illuminated solar cell," in: Rec. Tenth IEEE Photovoltaic Spec. Conf., Palo Alto, Cal., 1973, IEEE, pp. 188-193, New York, 1974.

66. T. B. S. Chadda and M. Wolf, "Comparison of vertical multi-junction and conventional solar cell performance," in: Rec. Tenth IEEE Photovoltaic Spec. Conf., Palo Alto, Cal., 1973, IEEE, pp. 52-57, New York, 1974.

67. M. B. Prince, "Silicon solar energy converters," J. Appl. Phys., vol. 26, pp. 534-540, 1955.

68. M. Wolf and H. Rauschenbach, "Series resistance effects on solar cell measurements," Adv. Energy Convers., vol. 3, pp. 455-479, 1963.

69. W. Shockley, "The theory of p–n junctions in semiconductors and p–n junction transistors," Bell Syst. Tech. J., vol. 28, pp. 435-489, 1949.

70. C. T. Sah, R. N. Noyce, and W. Shockley, "Carrier generation in p–n junctions and p–n junction characteristics," Proc. IRE, vol. 45, pp. 1228-1243, 1957.

71. V. M. Evdokimov, "Calculation of the series and parallel resistances from the current-voltage characteristic of a solar cell," Geliotekhnika [Applied Solar Energy], no. 6, pp. 16-22, 1972.

72. B. Ross and I. R. Madigan, "Thermal generation of recombi-

nation centers in silicon," Phys. Rev., vol. 108, pp. 1428-1433, 1957.

73. R. Gereth, H. Fischer, E. Link, et al., "Contribution to silicon solar cell technology," Energy Convers., vol. 12, pp. 103-107, 1972.

74. K. Graff and H. Fischer, "The lifetime of carriers in silicon and its influence on the characteristics of solar cells," in: Solar Energy Conversion: Solid State Physics [Russian translation], M. M. Koltun and V. M. Evdokimov (Editors), Energoizdat, 1982.

75. N. M. Bordina, A. M. Vasil'ev, A. K. Zaitseva, and A. P. Landsman, "Effect of spreading resistance on the load characteristic of silicon photocells with different current contacts," Radiotekh. Elektron., vol. 10, pp. 727-735, 1965.

76. N. M. Bordina and A. K. Zaitseva, "Choice of the optimum dimensions and load for a silicon solar cell with different current leads," Radiotekh. Elektron., vol. 10, pp. 1356-1358, 1965.

77. J. Mandelkorn, J. Lamneck, and R. L. Scrudder, "Design, fabrication and characteristics of new types of back surface field cell," in: Rec. Tenth IEEE Photovoltaic Spec. Conf., Palo Alto, Cal., 1973, IEEE, pp. 207-211, New York, 1974.

78. J. Lindmayer and J. F. Allison, British Patent No. 14116097, Solar Cell with Ta_2O_5 Antireflective Coating, Application No. 25666/73, 30 May 1973; published 3 December 1975.

79. J. Lindmayer and J. F. Allison, "The violet cell: an improved silicon solar cell," Comsat Tech. Rev., vol. 3, pp. 1-22, 1973.

80. V. M. Evdokimov, "Effect of internal drift fields on the efficiency of a semiconductor drift photocell and a graded gap cell," Radiotekh. Elektron., vol. 10, pp. 1314-1324, 1965.

81. R. L. Cummerow, "The photoeffect in p—n junction," Phys. Rev., vol. 95, pp. 16-21, 1954.

82. J. J. Loferski, "Theoretical considerations governing the choice of the optimum semiconductor for photovoltaic solar energy conversion," J. Appl. Phys., vol. 27, pp. 777-784, 1956.

83. C. G. Abbot, The Sun [Russian translation by Prof. E. Ya. Perepelkin], ONTI, Moscow-Leningrad, 1936.

84. M. Wolf, "Drift fields in photovoltaic solar energy converter cells," Proc. IEEE, vol. 51, pp. 674-693, 1963.

85. G. L. Bir and G. E. Pikus, "Effect of surface recombination on the efficiency of a photocell with a p—n junction," Zh. Tekh. Fiz., vol. 27, pp. 467-472, 1957.

86. J. J. Loferski and J. J. Wysocki, "Spectral response of photovoltaic cells," RCA Rev., vol. 22, pp. 38-56, 1961.

87. G. C. Jain and R. M. S. Al-Rifai, "Effects of drift fields and field gradients on the quantum efficiency of photocells," J. Appl. Phys., vol. 38, pp. 768-774, 1967.

88. V. K. Subashiev, "Determination of the recombinational constants from the spectral parameters of a photocell with a p—n junction," Fiz. Tverd. Tela (Leningrad), vol. 2, pp. 205-212, 1960.

89. V. K. Subashiev, G. B. Dubrovskii, and V. A. Petrusevich, "Determination of recombinational constants and depth of p—n junction from the spectral characteristics of photocells," Fiz. Tverd. Tela (Leningrad), vol. 2, pp. 1978-1980, 1960.

90. G. B. Dubrovskii, "Determination of the basic parameters of photocells with a p—n junction from the spectral characteristics," Zavod. Lab., vol. 27, pp. 1233-1236, 1961.

91. J. A. Baicker and B. W. Faughnan, "Radiation-induced changes in silicon photovoltaic cells," J. Appl. Phys., vol. 33, pp. 3271-3280, 1962.

92. N. M. Bordina and T. M. Golovner, "Determination of the diffusion length of minority carriers in the base of silicon photoconverters," Geliotekhnika [Applied Solar Energy], no. 1, pp. 11-16, 1977.

93. J. H. Reynolds and A. Meulenberg, Jr., "Measurement of diffusion length in solar cells," J. Appl. Phys., vol. 45, pp. 2582-2592, 1974.

94. J. J. Wysocki, P. Rappaport, E. Davison, and J. J. Loferski, "Low-energy proton bombardment of GaAs and Si solar cells," IEEE Trans. Electron Devices, vol. ED-13, pp. 420-429, 1966.

95. W. Rozenzweig, "Radiation effects in silicon devices," IEEE Trans. Nucl. Sci., vol. 12, pp. 18-19, 1965.

96. D. J. Curtin and R. L. Statler, "Review of radiation damage

to silicon solar cells," IEEE Trans. Aerospace Electron System, vol. 11, pp. 499-513, 1975.

97. E. B. Vinogradova, T. M. Golovner, S. M. Gorodetskii, G. M. Grigor'eva, E. V. Zhidkova, A. E. Zaitseva, and L. B. Kreinin, "Spectral characteristics of photoconverters with a nonuniform distribution of defects in the base," Geliotekhnika [Applied Solar Energy], no. 1, pp. 13-17, 1978.

98. M. B. Kagan and T. L. Lyubashevskaya, "Determination of recombinational parameters and depth of the p—n junction in semiconductor photocells," Geliotekhnika [Applied Solar Energy], no. 4, pp. 11-15, 1968.

99. V. M. Evdokimov, "Determination of the parameters of minority carriers in semiconductor photocells using the spectral sensitivity curve," Geliotekhnika [Applied Solar Energy], no. 3, pp. 32-38, 1972.

100. R. Gremmelmaier, "Irradiation of p—n junctions with gamma-rays," Proc. IRE, vol. 46, pp. 1045-1049, 1958.

101. W. Rozenzweig, "Diffusion length measurement by means of ionizing radiation," Bell Syst. Tech. J., vol. 41, pp. 1573-1589, 1962.

102. M. I. Epshtein, Spectral Measurements in Electrovacuum Technology, Energiya, Moscow, 1970.

103. Z. M. Dashevskii, M. L. Dobrokhotova, N. L. Emel'yanova, I. A. Krasnopol'skaya, I. Ya. Ravich, I. V. Sbignev, and T. N. Toroptseva, "Properties of dielectric substrates for highly sensitive film thermoelectric sensors," Elektrotekh. Prom-st. Ser. Fiz. Khim. Istochniki Toka, no. 4, pp. 3-5, 1982.

104. I. M. Vesel'nitskii, Yu. D. Ignat'ev, A. S. Il'in, and A. B. Fromberg, "The thermocouples RTN-16S, RTN-16G, RTN-30, and RTN-30G and their metrological applications," in: Photometry and Its Metrological Applications. Abstracts of Papers read to the Third All-Union Scientific and Technological Conference, 2-6 December 1979, Moscow, VNIIOFI, p. 49, Moscow, 1980.

105. E. I. Agaev, M. M. Koltun, and A. L. Kostanenko, "Radiative power of the optical surfaces of solar energy converters investigated with a high-precision radiometer," in: Solar Energy Conversion. Abstracts of Papers read to the Chernogolovka

Conf., 17-19 February 1981, Institute of Chemical Physics of the USSR Academy of Sciences, pp. 161-162, 1981.

106. M. B. Kagan, A. P. Landsman, and Ya. I. Chernov, "Analysis of the spectral and temperature characteristics of photoelectric converters and the choice of effective applications," Kosmich. Issled., vol. 4, pp. 128-136, 1966.

107. M. Wolf, "A new look at silicon solar cell performance," Energy Convers., vol. 11, pp. 63-73, 1971.

108. M. B. Kagan, M. M. Koltun, A. P. Landsman, and T. L. Lyubashevskaya, "Possible designs for cascade solar cells," Geliotekhnika [Applied Solar Energy], no. 1, pp. 7-20, 1968.

109. M. M. Koltun and A. P. Landsman, "Possibility of the ideal spectral distribution of the reflection coefficient of silicon photocells," Opt. Spektrosk., vol. 26, pp. 618-621, 1969.

110. M. M. Koltun and A. P. Landsman, "Semiconductor photoconverters transparent in the solar infrared," Zh. Prikl. Spektrosk., vol. 15, no. 753-755, 1971.

111. M. M. Koltun, V. P. Matveev, and V. A. Unishkov, "Semiconductor photocells transparent in the long-wave part of the spectrum beyond the fundamental absorption band," Geliotekhnika [Applied Solar energy], no. 5, pp. 10-17, 1973.

112. S. M. Bedair and M. P. Lamorte, "A two-junction cascade solar-cell structure," Appl. Phys. Lett., vol. 34, pp. 38-39, 1979.

113. M. B. Kagan and T. L. Lyubashevskaya, "New designs for and optimization of cascade solar cells," Geliotekhnika [Applied Solar Energy], no. 6, pp. 7-15, 1981.

114. R. L. Moon, L. W. James, H. A. Van der Plas, et al., "'Multigap solar cell requirements and the performance of AlGaAs and Si cells in concentrated sunlight," in: Rec. Thirteenth IEEE Photovolt. Spec. Conf., Wash., 1979, pp. 859-867, IEEE, New York, 1978.

115. Zh. I. Alferov, V. M. Andreev, D. Z. Garbuzov, and V. D. Rumyantsev, "The 100% radiative-recombination internal quantum yield in three-layer lightguides based on AlAs–GaAs," Fiz. Tekh. Poluprovodn., vol. 9, pp. 462-469, 1975.

116. Zh. I. Alferov, V. M. Andreev, M. B. Kagan, I. I. Protasov, and V. T. Trofim, "Solar converters based on p-$Ga_{1-x}Al_x$ As–

n-GaAs heterojunctions," Fiz. Tekh. Poluprovodn., vol. 4, pp. 2378-2379, 1970.

117. Zh. I. Alferov, V. M. Andreev, G. S. Daletskii, M. B. Kagan, N. S. Lidorenko, and V. M. Tuchkevich, "High-efficiency heterophotoconverters using the GaAs—AlAs system," Proc. World Electrotechnical Congress, 21-25 June 1977, Moscow, Sect. 5A, 04, Informelektro, Moscow, 1977.

118. V. K. Subashiev and E. M. Pedyash, "Energy diagram of a real silicon photocell," Fiz. Tverd. Tela (Leningrad), vol. 2, pp. 213-220, 1960.

119. A. M. Vasil'ev, "Effect of the drift field on the collection coefficient in a silicon photocell," Geliotekhnika [Applied Solar Energy], no. 3, pp. 7-12, 1967.

120. V. M. Evdokimov, "Photocurrent in solar cells with an inhomogeneous built-in field," Geliotekhnika [Applied Solar Energy], no. 5, pp. 3-20, 1972.

121. I. P. Gavrilova, "Comparison of the efficiency of photocells with stepped and exponential impurity distributions in the doped layer," Geliotekhnika [Applied Solar Energy], no. 6, pp. 23-28, 1972.

122. I. P. Gavrilova, V. M. Evdokimov, M. M. Koltun, V. P. Matveev, and E. S. Makarova, "Enhanced efficiency semiconductor photocells with drift field in the doped region," Fiz. Tekh. Poluprovodn., vol. 8, pp. 119-124, 1974.

123. G. E. Pikus, Fundamentals of the Theory of Semiconductor Devices, pp. 369-370, Nauka, Moscow, 1965.

124. K. Bachman, "Materials for solar cells," in: Current Problems in Materials Science [Russian translation], E. I. Givargizov and M. M. Koltun (Editors), no. 1, pp. 7-195, Mir, Moscow, 1982.

125. M. M. Koltun, E. S. Makarova, and I. P. Gavrilova, "Impurity distribution in the silicon surface layer due to the diffusion of phosphorus through a porous oxide film," Izv. Akad. Nauk SSSR. Neorg. Mater., vol. 12, pp. 161-165, 1976.

126. I. A. Minucci, K. W. Matthei, A. R. Kirkpatrick, and A. McGrosky, "Silicon solar cells with high open-circuit voltage," in: Rec. Fourteenth IEEE Photovolt. Spec. Conf., San Diego, Cal., 1980, pp. 93-96, IEEE, New York, 1980.

127. R. D. Nasby and J. C. Fossum, "Characterization of p^+-n-n^+

BSF silicon concentrator solar cells," in: Rec. Fourteenth IEEE Photovolt. Spec. Conf., San Diego, Cal., 1980, pp. 419-422, IEEE, New York, 1980.

128. R. Van Overstraeten and W. Nuyts, "Theoretical investigation of the efficiency of drift-field solar cells," IEEE Trans. Electron. Devices, vol. ED-16, pp. 632-641, 1969.

129. A. Usami and M. Yamaguchi, "Lithium-doped drift field radiation-resistant p—n-type silicon solar cells," in: Rec. Eleventh IEEE Photovolt. Spec. Conf., Scottsdale, Ariz., 1975, pp. 13-18, IEEE, New York, 1975.

130. C. R. Baraona and H. W. Brandhorst, "Analysis of epitaxial drift field n on p silicon solar cells," in: Rec. Twelfth IEEE Photovolt. Spec. Conf., Baton Rouge, La., 1976, pp. 9-14, IEEE, New York, 1976.

131. G. A. Chetverikova, "Si-Base photoconverters with a drift field in the base," Geliotekhnika [Applied Solar Energy], no. 1, pp. 15-18, 1976.

132. M. M. Koltun, V. V. Arsenin, and B. M. Abdurakhmanov, "Silicon solar cells based on epitaxial structure," Geliotekhnika [Applied Solar Energy], no. 5, pp. 87-91, 1981.

133. V. S. Vavilov and N. A. Ukhin, Radiation Effects in Semiconductors and Semiconductor Devices, Atomizdat, Moscow, 1969.

134. I. I. Wysocki, "Role of lithium in damage and recovery of irradiated silicon solar cells," IEEE Trans. Nucl. Sci., vol. 13, pp. 168-173, 1966.

135. J. Mandelkorn and J. H. Lamneck, "New electric field effect in silicon solar cells," J. Appl. Phys., vol. 44, pp. 4785-4791, 1973.

136. P. Iles, "Increased output from silicon solar cells," in: Rec. Eighth IEEE Photovolt. Spec. Conf., Seattle, Wash., 1970, pp. 345-352, IEEE, New York, 1970.

137. K. S. Azimov, N. M. Bordina, G. M. Grigor'eva, L. B. Kreinin, and A. P. Landsman, "Photoelectric characteristics of photoconductors made from high-resistivity silicon," Fiz. Tekh. Poluprovodn., vol. 7, pp. 2257-2260, 1973.

138. W. Palz, J. Besson, T. Nguyen Duy, and J. Vedel, "Review of CdS solar cell activities," in: Rec. Tenth IEEE Photovolt.

Spec. Conf., Palo Alto, Cal., 1973, pp. 69-76, IEEE, New York, 1974.

139. A. Meulenberg, D. J. Curtin, and R. W. Cool, "Comparative testing of high efficiency solar cells," in: Rec. Twelfth IEEE Photovolt. Spec. Conf., Baton Rouge, La., 1976, pp. 238-246, IEEE, New York, 1976.

140. T. M. Golovner, A. A. Lebedeva, N. V. Penkina, and N. K. Sirotenko, "Effect of the thickness of the titanium layer on the reflective power of silicon photoconverters in the near-infrared," Geliotekhnika [Applied Solar Energy], no. 2, pp. 7-18, 1982.

141. M. M. Koltun and V. P. Matveev, "Semiconductor photoconverters with heat-reflecting contacts," in: Conversion and Utilization of Solar Energy, G. M. Krzhizhanovskii Power Institute, no. 6, pp. 61-67, Moscow, 1973.

142. G. S. Daletskii, M. B. Kagan, M. M. Koltun, and V. M. Kuznetsov, "Development of solar batteries for the Venera-9 and Venera-10 automatic interplanetary stations and the Lunokhod program," Geliotekhnika [Applied Solar Energy], no. 4, pp. 3-9, 1979.

143. N. M. Bordina, N. A. Borisova, G. S. Daletskii, V. V. Zadde, A. K. Zaitseva, A. P. Landsman, and V. A. Letin, "Using the radiation reflected from the Earth to increase the power of solar batteries," Kosmich. Issled., vol. 14, pp. 293-299, 1976.

144. A. K. Zaitseva and O. P. Fedoseeva, "Possible application of two-sided silicon photoconverters," in: Semiconductor Solar Energy Converters, USSR Academy of Sciences, pp. 87-90, Moscow, 1961.

145. J. J. Capart, "Design of solar cells with two collecting junctions," Europ. Space Res. Organization (ESRO), Techn. Note TN-3 (ESTEC), p. 25, November, 1966.

146. N. M. Bordina, T. M. Golovner, V. V. Zadde, A. K. Zaitseva, A. P. Landsman, and V. I. Strel'tsova, "Operation of a thin silicon photoconverter illuminated from both sides," Geliotekhnika [Applied Solar Energy], no. 6, pp. 10-19, 1975.

147. Y. Chevalier, F. Duenas, and J. Chambouleyron, "A novel photovoltaic panel for low cost power," in: Rec. Thirteenth IEEE Photovolt. Spec. Conf., Wash., 1978, pp. 738-743, IEEE, New York, 1978.

148. N. B. Rekant and A. V. Sheklein, "Calculations of the integrated optical parameters of materials for solar radiation applications," in: Low-power solar systems: Proc. First Seminar on Solar Energy, Moscow, 29 June–13 July 1979, G. M. Krzhizhanovskii Power Institute, pp. 127-139, Moscow, 1980.

149. G. A. Boltyanskii, N. M. Bordina, G. S. Daletskii, V. G. Ermakov, V. R. Zayavlin, V. A. Letin, and M. N. Kholeva, "The experimental two-sided solar battery for the Salyut-5 orbital station," Kosmich. Issled., vol. 18, pp. 812-814, 1980.

150. D. A. Jenny, J. J. Loferski, and P. Rappaport, "Photovoltaic effect in GaAs p–n junctions and solar energy conversion," Phys. Rev., vol. 101, pp. 1208-1209, 1956.

151. R. Gremmelmaier, "GaAs photocells," Z. Naturforsch. Part A, vol. 10, pp. 501-507, 1955.

152. G. W. Turner, J. C. Fan, R. L. Chapman, and R. P. Gale, "GaAs shallow homojunction concentrator solar cells," in: Rec. Fifteenth IEEE Photovolt. Spec. Conf., Kissimee, Fla., 1981, pp. 151-155, IEEE, New York, 1981.

153. O. Madelung, Physics of Three-Five Compounds, Meyerhoffer, Dietrick, 1964 [Russian translation], B. I. Boltaks (Editor), Mir, Moscow, 1967.

154. M. B. Kagan, A. P. Landsman, and Ya. I. Chernov, "A photocell with extended spectral sensitivity," Prib. Tekh. Eksp., vol. 19, pp. 232-233, 1965.

155. Zh. I. Alferov, V. M. Andreev, N. S. Zimogorova, and D. N. Tret'yakov, "Photoelectric properties of the heterojunctions $Al_xGa_{1-x}As$–GaAs," Fiz. Tekh. Poluprovodn., vol. 3, pp. 1633-1637, 1969.

156. H. J. Hovel, "Solar cells for terrestrial application," Solar Energy, vol. 19, pp. 605-615, 1977.

157. J. M. Woodall and H. J. Hovel, "High efficiency $Ga_{1-x}Al_xAs$–GaAs solar cells," App. Phys. Lett., vol. 2I(1), pp. 379-381, 1972.

158. H. J. Hovel and J. M. Woodall, "Theoretical and experimental evaluations of $Ga_{1-x}Al_xAs$–GaAs solar cells," in: Rec. Tenth IEEE Photovolt. Spec. Conf., Palo Alto, Cal., 1973, pp. 25-30, IEEE, New York, 1974.

159. H. J. Hovel and J. M. Woodall, "$Ga_{1-x}Al_xAs$–GaAs p–p–n

heterojunction solar cells," J. Electrochem. Soc., vol. 120, pp. 1246-1251, 1973.

160. V. M. Andreev, M. B. Kagan, T. L. Lyubashevskaya, T. A. Nuller, and D. N. Tret'yakov, "Comparison of different heteroconverters based on the p-$Al_xGa_{1-x}As$–n-GaAs system in terms of maximum possible efficiency," Fiz. Tekh. Poluprovodn., vol. 8, pp. 1328-1334, 1974.

161. G. S. Kamath, I. Ewan, and R. C. Knechtli, "High efficiency and large area GaAl/As–GaAs solar cells," in: Rec. Twelfth IEEE Photovolt. Spec. Conf., Baton Rouge, La., 1976, pp. 929-933, IEEE, New York, 1976.

162. W. D. Johnston and W. M. Callahan, "Vapor-phase-epitaxial growth, processing and performance of AlAs–GaAs heterojunction solar cells," in: Rec. Twelfth IEEE Photovolt. Spec. Conf., Newark, Del., 1975, IEEE, pp. 396–399, New York, 1975.

163. G. S. Daletskii, M. B. Kagan, N. S. Lidorenko, and S. V. Ryabikov, "Photoelectric converters and solar batteries based on complex semiconducting materials and heterostructures," in: Solar Energy Conversion: Abstracts of Papers read to the Chernogolovka Conf., 17-19 February 1981, Institute of Chemical Physics of the USSR Academy of Sciences, pp. 62-63, Chernogolovka, 1981.

164. Zh. I. Alferov, V. M. Andreev, Yu. M. Zadiranov, V. I. Korol'kov, and T. S. Tabarov, "Converters of concentrated solar radiation using continuous and lumped Al–Ga–As heterostructures," in: Solar Energy Conversion: Abstracts of Papers read to the Chernogolovka Conf., 17-19 February 1981, Institute of Chemical Physics of the USSR Academy of Sciences, pp. 10-11, Chernogolovka, 1981.

165. H. C. Casey and M. B. Panish, "Composition dependence of the $Ga_{1-x}Al_xAs$ direct and indirect energy gap," J. Appl. Phys., vol. 40, pp. 4910-4912, 1969.

166. Zh. I. Alferov and V. M. Andreev, "Prospects for the photoelectric method of solar energy conversion," in: Solar Energy Conversion, N. N. Semenov (Editor), Institute of Chemical Physics, USSR Academy of Sciences, pp. 7-20, Chernogolovka, 1981.

167. R. Sahai, D. D. Edwall, E. Cory, and J. S. Harris, "High ef-

ficiency thin window $Ga_{1-x}Al_xAs$—GaAs solar cells," in: Rec. Twelfth IEEE Photovolt. Spec. Conf., Baton Rouge, La., 1976, pp. 989-992, IEEE, New York, 1976.

168. A. N. Imenkov, A. A. Stamkulov, T. I. Taurbaev, B. V. Tsarenkov, V. F. Shorin, and Yu. P. Yakovlev, "High-efficiency solar photoconverters with a thin graded-gap layer," Fiz. Tekh. Poluprovodn., vol. 12, pp. 948-951, 1978.

169. I. P. Kalinkin, V. B. Aleskovskii, and A. V. Simashkevich, Epitaxial Films of $A^{II}B^{VI}$ Compounds, Leningrad State University, 1978.

170. A. V. Simashkevich, Heterojunctions Based on Semiconducting Compounds $A^{II}B^{VI}$, Shtiintsa, Kishinev, 1980.

171. D. C. Reynolds, C. Leies, L. L. Antes, and R. E. Marburger, "Photovoltaic effect in cadmium sulfide," Phys. Rev., vol. 96, pp. 533-534, 1954.

172. F. A. Shirland, "The history, design, fabrication and performance of CdS thin film solar cells," Adv. Energy Convers., vol. 6, pp. 201-221, 1966.

173. L. R. Shiozawa, G. A. Sullivan, and F. Augustine, "The mechanism of the photovoltaic effect in high efficiency CdS thin-film solar cells," in: Seventh IEEE Photovolt. Spec. Conf., Pasadena, Cal., 1968, pp. 39-46, IEEE, New York, 1968.

174. T. S. Te Velde, "The production of the cadmium sulfide-copper sulfide solar cell by means of a solid-state reaction," Energy Convers., vol. 14, pp. 111-115, 1975.

175. L. S. Burton and G. Haake, "New type Cu_2S/CdS backwall solar cell," in: Rec. Tenth IEEE Intersoc. Energy Conversion Conf., Newark, De., 1975, IEEE, pp. 396—399, New York, 1975.

176. A. Rothwarf and A. M. Barnett, "Design analysis of the thin-film CdS—Cu_2S solar cell," IEEE Trans. Electron. Devices, vol. ED-24, pp. 381-387, 1977.

177. I. V. Egorova, "Photovoltaic effect in the CdS—Cu_2S system," Fiz. Tekh. Poluprovodn., vol. 2, pp. 319—323, 1968.

178. I. V. Karpenko, M. M. Koltun, and R. N. Tykvenko, "Prospects and problems in the development of thin-film solar cells based on $A^{II}B^{IV}$ compounds," in: Solar Energy Conversion: Abstracts of Papers read at the Chernogolovka Conf.

on 17-19 February 1981, Institute of Chemical Physics, USSR Academy of Sciences, pp. 39-40, Chernogolovka, 1981.

179. A. Smith, "Thin-film CdS/Cu_2S heterojunctions: dark $I-V$ characteristics and heat treatment," J. Appl. Phys., vol. 50, pp. 1160-1162, 1979.

180. R. V. Kantariya and S. Yu. Pavelets, "Energy band diagram for the $p\text{-}Cu_{2-x}S\text{-}n\text{-}CdS$ heterojunction," Fiz. Tekh. Poluprovodn., vol. 12, pp. 1214-1217, 1978.

181. V. N. Komashchenko, A. I. Marchenko, and G. A. Fedorus, "Thin-film and ceramic solar converters based on cadmium sulfide and selenide," Geliotekhnika [Applied Solar Energy], no. 3, pp. 15-21, 1979.

182. J. Vedel and E. Castel, "Electrochemical determination of the copper sulfide composition: application to $CdS-Cu_2S$ solar cells," in: Proc. Intern. Congr. "The Sun at the service of mankind"; The Section "Photovoltaic power and its application in space and on Earth," 2-6 July 1973, Cent. Nat. Etudes Spatiales, pp. 199-205, Bretigny-sur-Orge, 1973.

183. M. Savelli and J. Bugnot, "Problems in developing photocells based on Cu_2S/CdS," in: Solar Energy Converters [Russian translation], M. M. Koltun and V. M. Evdokimov (Editors), pp. 189-226, Energoizdat, Moscow, 1982.

184. B. J. Mulder, "Optical properties of crystals of cuprous sulfides (chalcocite, djurleite, $Cu_{1.9}S$ and digenite)," Phys. Status Solidi (A), vol. 13, pp. 79-88, 1972.

185. E. M. Voronkova, B. N. Grechushnikov, G. I. Distler, and I. P. Petrov, Optical Materials for Infrared Technology, Nauka, Moscow, 1965.

186. J. A. Bragagnolo, R. W. Birkmire, and J. E. Phillips, "Thin-film CdS/Cu_2S with high open-circuit voltage and low reflection losses," in: Rec. Fourteenth IEEE Photovolt. Spec. Conf., San Diego, Cal., 1980, pp. 1400-1401, IEEE, New York, 1980.

187. M. M. Koltun and M. A. Razykova, "Electric and photoelectric properties of $Cu_2S-Zn_xCd_{1-x}S$ heterojunctions," Geliotekhnika [Applied Solar Energy], no. 2, pp. 15-22, 1982.

188. J. J. Loferski, J. Shewchun, E. A. Demeo, et al., "Compari-

son of some properties of Cu_2S-CdS photovoltaic cells, in which different methods are used to prepare the (Cu_2S) layer," in: Proc. Intern. Conf. on Solar Electricity, Toulouse, France, 1-5 March 1976, Cent. Nat. Etudes Spatiales, pp. 317-324, Bretigny-sur-Orge, 1976.

189. L. L. Kazmerski, F. R. White, and G. K. Morgan, "Thin-film $CuInSe_2-CdS$ solar cell," Appl. Phys. Lett., vol. 29, pp. 268-270, 1976.

190. K. J. Bachman, E. Buehler, J. G. Shay, and S. Wagner, "Polycrystalline thin-film $InP-CdS$ solar cell," Appl. Phys. Lett., vol. 29, pp. 121-123, 1976.

191. E. M. Konstantinova, N. R. Stratieva, S. K. Kynev, and L. V. Vasil'ev, "Photocells based on heterojunctions involving polycrystalline cadmium sulfide and chromium telluride," Geliotekhnika [Applied Solar Energy], no. 2, pp. 23-25, 1982.

192. M. I. Krunks, E. Ya. Mellikov, P. L. Kukk, I. V. Karpenko, and K. N. Puchkova, "Chemically pulverized CdS and CdZnS films for solar energy converters," in: Solar Energy Conversion: Abstracts of Papers read at the Chernogolovka Conf. on 17-19 February 1981, Institute of Chemical Physics, USSR Academy of Sciences, pp. 44-45, Chernogolovka, 1981.

193. Kh. T. Akramov, G. Ya. Umarov, and T. M. Razykov, "Effect of thermal treatment on certain properties of the thin-film Cu_xS-CdS heterojunction," Geliotekhnika [Applied Solar Energy], no. 6, pp. 3-7, 1975.

194. S. A. Azimov, Sh. A. Mirsagatov, and D. T. Rasulov, "Photoelectric properties of thin-film p-CdTe-n-CdS heterojunctions," Geliotekhnika [Applied Solar Energy], no. 3, pp. 18-24, 1978.

195. D. E. Carlson, "An overview of amorphous silicon solar cell development," in: Rec. Fourteenth IEEE Photovolt. Spec. Conf., San Diego, Cal., 1980, pp. 291-297, IEEE, New York, 1980.

196. A. Madan, S. R. Ovshinsky, W. Czubatyi, and M. Shur, "Some electrical and optical properties of $\alpha = Si:F:H$ alloys," J. Electron. Mater., vol. 9, pp. 385-407, 1980.

197. D. E. Carlson, "Amorphous silicon cell development," in:

Proc. Photovolt. Adv. and Res. Ann. Rev. Meet., Denver, Col., 17-19 September 1979, pp. 113-119, SERI, Golden, Col., 1979.

198. Y. Hamakawa, H. Okamoto, and Y. Nitta, "Optimum design and device physics of the horizontally multilayered high voltage solar cells produced by plasma deposited amorphous silicon," in: Rec. Fourteenth IEEE Photovolt. Spec. Conf., San Diego, Cal., 1980, pp. 1074-1079, IEEE, New York, 1980.

199. D. E. Carlson, C. R. Wronsky, A. R. Triano, and R. E. Daniel, "Solar cells using Schottky barriers on amorphous silicon," in: Rec. Twelfth IEEE Photovolt. Spec. Conf., Baton Rouge, La., 1976, pp. 893-897, IEEE, New York, 1976.

200. B. Y. Tong, P. K. John, S. K. Wong, and K. R. Chik, "Highly stable, photosensitive evaporated amorphous silicon films," Appl. Phys. Lett., vol. 38, pp. 789-790, 1981.

201. V. G. Litovchenko and V. G. Popov, "An investigation of the properties of hydrogenized amorphous silicon investigated by the method of spectral characteristics of capacitor photo-emf," Fiz. Tekh. Poluprovodn., vol. 16, pp. 734-738, 1982.

202. J. I. Pankove, C. P. Wu, G. W. Magee, and J. T. McGinn, "Laser annealing of hydrogenated amorphous silicon," J. Electron. Mater., vol. 99, pp. 905-912, 1980.

203. I. P. Akimchenko, V. S. Vavilov, and N. N. Dymova, "Suppression of the Staebler-Wronski effect in amorphous hydrogenized silicon under gallium and arsenic ion implantation," Pis'ma Zh. Eksp. Teor. Fiz., vol. 33, pp. 448-451, 1980.

204. H. Okamoto, Y. Nitta, T. Adachi, and Y. Hamakawa, "Glow discharge produced amorphous silicon solar cells," Surface Sci., vol. 86, pp. 486-491, 1979.

205. D. E. Carlson, "Amorphous thin films for terrestrial solar cells," Proc. Twenty-eighth Nat. Symp. Amer. Vacuum Soc., Anaheim, Cal., 2-6 November 1981, J. Vac. Sci. Technol., Pt. 1, vol. 20, pp. 290-295, 1982.

206. J. Scott-Monck and P. Stella, "Factors governing photovoltaic technology transfer: terrestrial to space," J. Energy, vol. 6, pp. 16-19, 1982.

207. R. T. Young, R. F. Wood, and W. H. Christie, "Laser proces-

sing for high-efficiency Si solar cells," J. Appl. Phys., vol. 53, pp. 1178-1189, 1982.

208. M. B. Kagan, N. S. Koroleva, and T. A. Nuller, "Solar cells based on epitaxial gallium arsenide films," Geliotekhnika [Applied Solar Energy], no. 2, pp. 28-32, 1970.

209. N. M. Bordina, M. B. Kagan, V. A. Letin, T. L. Lyubashevskaya, and T. A. Nuller, "A study of the parameters of GaAlAs—GaAs solar cells designed for operation under two-sided illumination," in: Solar Energy Conversion: Abstracts of Papers read at the Chernogolovka Conf. on 17-19 February 1981, Institute of Chemical Physics, USSR Academy of Sciences, pp. 16-17, Chernogolovka, 1981.

210. R. P. Gale, B. Y. Tsaur, J. C. Fan, et al., "GaAs shallow-homojunction solar cells on epitaxial Ge grown on Si substrates," in: Rec. Fifteenth IEEE Photovoltaic Spec. Conf., Kissimee, Fla., 1981, pp. 1051-1055, IEEE, New York, 1981.

211. V. A. Benderskii, M. I. Fedorov, and N. N. Usov, "Quantum yield of the barrier photoelectric effect in phthalocyanine layers," Dokl. Akad. Nauk SSSR, vol. 183, pp. 1117-1119, 1968.

212. R. O. Louffy, I. H. Sharp, C. K. Hsiao, and R. Ho, "Phthalocyanine organic solar cells: indium/X-metal free phthalocyanine Schottky barriers," J. Appl. Phys., vol. 52, pp. 5218-5230, 1981.

213. A. Yu. Borisov, "Photosynthesis — a possible way to solar power engineering," in: Solar Energy Conversion, N. N. Semenov (Editor), Institute of Chemical Physics, USSR Academy of Sciences, pp. 138-142, Chernogolovka, 1981.

214. H. Gerischer, "Photoelectrolysis under the influence of solar radiation using semiconductor electrodes," in: Solar Energy Conversion [Russian translation], M. M. Koltun and V. M. Evdokimov (Editors), pp. 106-150, Energoizdat, Moscow, 1982.

215. A. G. Milnes and A. G. Feucht, Heterojunctions and Metal Semiconductor Junctions [Russian translation], V. S. Vavilov (Editor), Mir, Moscow, 1975.

216. V. I. Strikha, Theoretical Principles of the Metal Semiconductor Contact, Naukova Dumka, Kiev, 1974.

217. E. Farb and R. Tizhburg, "Solar cells using the In_2O_3 (n^+) heterojunction," in: Solar Power Engineering [Russian translation], Yu. N. Malevskii and M. M. Koltun (Editors), pp. 261-266, Mir, Moscow, 1979.

218. O. P. Agnihotri and B. K. Gunta, Solar Selective Surfaces [Russian translation], M. M. Koltun (Editor), Mir, Moscow, 1984.

219. W. A. Anderson, A. E. Delahoy, and R. A. Milano, "Thin metal films as applied to Schottky solar cells: optical studies," Appl. Opt., vol. 15, pp. 1621-1625, 1976.

220. V. V. Arsenits, M. M. Koltun, and A. I. Kulagin, "Optical characteristics of solar cells based on the metal—oxide—semiconductor structure," Geliotekhnika [Applied Solar Energy], no. 3, pp. 72-74, 1980.

221. A. I. Malik, V. A. Manasson, and V. A. Baranyuk, "Role of tunnel-thin dielectric in MOS solar converters," Geliotekhnika [Applied Solar Energy], no. 3, pp. 72-74, 1981.

222. O. S. Zinets, S. S. Kil'chitskaya, and V. I. Strikha, "Effect of the intermediate layer on the characteristics of Schottky barrier solar cells," Geliotekhnika [Applied Solar Energy], no. 1, pp. 15-19, 1982.

223. A. I. Malik, V. A. Baranyuk, and V. A. Manasson, "Improved solar converters based on the In_2O_3/SnO_2-SiO_x-n-Si structure," Geliotekhnika [Applied Solar Energy], no. 1, pp. 3-4, 1980.

224. Em. Saucedo and I. Mimila-Arroyo, "A 14% efficiency SnO_x-SiO_2-n-Si solar cell," in: Rec. Fourteenth IEEE Photovolt. Spec. Conf., San Diego, Cal., 1980, pp. 1370-1375, IEEE, New York, 1980.

225. B. I. Gil'man, V. V. Kasatkin, Yu. V. Sorokin, Yu. V. Skokov, and M. G. Zaks, "Characteristics of silicon photoconverters with an inversion layer," Geliotekhnika [Applied Solar Energy], no. 3, pp. 31-38, 1978.

226. H. J. Hovel, "Solar cell for terrestrial applications," Solar Energy, vol. 19, pp. 605-615, 1977.

227. Shang-Yi-Chiang, B. G. Garbajal, and G. F. Wakefield, "Thin tandem junction solar cell," in: Rec. Thirteenth Photovolt. Spec. Conf., Wash., 1978, pp. 1290-1293, IEEE, New York, 1978.

228. N. G. Tarr, D. L. Pulfrey, and P. A. Iles, "Induced back surface field and MISIM solar cells on p-Si substrates," in: Rec. Fifteenth Photovolt. Spec. Conf., Kissimee, Fla., 1981, pp. 1409-1411, IEEE, New York, 1981.

229. R. J. Stirn and V. C. M. Veh, "A 15% efficient antireflection-coated metal-oxide-semiconductor solar cell," Appl. Phys. Lett., vol. 27, pp. 95-98, 1975.

230. D. Burk, J. Shewchun, M. Spitzer, et al., "Semiconductor—insulator—semiconductor (SIS) solar cells: indium—tin—oxide on silicon," in: Rec. Fourteenth IEEE Photovolt. Spec. Conf., San Diego, Cal., 1980, pp. 1376-1383, IEEE, New York, 1980.

231. T. Feng, A. K. Ghosh, H. P. Maruska, and D. I. Eustace, "On stability of SnO_2/n-Si and ITO/n-Si solar cells," in: Rec. Fifteenth IEEE Photovolt. Spec. Conf., Kissimee, Fla., 1981, pp. 1412-1417, IEEE, New York, 1981.

232. G. M. Storti, "The fabrication of 17% AM1 efficient semicrystalline silicon solar cell," in: Rec. Fifteenth IEEE Photovolt. Spec. Conf., Kissimee, Fla., 1981, pp. 442-445, IEEE, New York, 1981.

233. P. R. Emtage, "Electrical conduction and the photovoltaic effect in semiconductors with position-dependent band gap," J. Appl. Phys., vol. 33, pp. 1950-1960, 1962.

234. N. S. Lidorenko and V. M. Evdokimov, "The photoelectric method of conversion: present state and possible future developments," in: Solar Energy Conversion, N. N. Semenov (Editor), Institute of Chemical Physics, USSR Academy of Sciences, pp. 20-27, Chernogolovka, 1981.

235. V. M. Evdokimov, Yu. L. Lisovskii, A. F. Milovanov, and D. S. Strebkov, "New theoretical models and prospects for increasing the efficiency of photoconverters," in: Solar Energy Conversion: Abstracts of Papers read at the Chernogolovka Conf. on 17-19 February 1981, Institute of Chemical Physics, USSR Academy of Sciences, pp. 68-69, Chernogolovka, 1981.

236. H. J. Hovel, "Novel materials and devices for sunlight concentrating systems," IBM J. Res. Dev., vol. 22, pp. 112-121, 1978.

237. Zh. I. Alferov, V. M. Andreev, Kh. K. Aripov, V. R.

Larionov, and V. D. Rumyantsev, "Model of a self-contained solar converter incorporating heterophotocells and concentrators," Geliotekhnika [Applied Solar Energy], no. 2, pp. 3-6, 1981.

238. G. E. Guazoni, "High temperature spectral emittance of oxides of erbium, samarium, neodymium and ytterbium," Appl. Spectrosc., vol. 26, pp. 60-65, 1972.

239. C. W. Kim and P. J. Schwartz, "A p—i—n thermophotovoltaic diode," IEEE Trans. Electron. Devices, vol. ED-16, pp. 657-663, 1969.

240. A. M. Vasil'ev, T. M. Golovner, A. P. Landsman, and N. S. Lidorenko, "Optical parameters of silicon photocells and the efficiency of the thermophotoelectric converter," Teplofiz. Vys. Temp., vol. 5, pp. 1079-1093, 1967.

241. Zh. I. Alferov, V. M. Andreev, D. Z. Garbuzov, and M. K. Trukan, "Efficient injection luminescence from electron-hole plasma in two heterojunction structures," Fiz. Tekh. Poluprovodn., vol. 8, pp. 561-565, 1974.

242. Zh. I. Alferov, V. M. Andreev, D. Z. Garbuzov, B. V. Egorov, V. R. Larionov, V. D. Rumyantsev, and O. M. Fedorova, "Highly efficient solar cells with intermediate converters, designed for operation with solar concentrators," Pis'ma Zh. Eksp. Teor. Fiz., vol. 4, pp. 1128-1130, 1978.

243. S. M. Bedair, S. B. Phatak, M. Timmons, et al., "Recent progress in the development of the cascade solar cell," in: Rec. Fourteenth IEEE Photovolt. Spec. Conf., San Diego, Cal., pp. 337-340, IEEE, New York, 1980.

244. D. L. Miller, S. W. Zehr, and J. S. Harris, "GaAs—AlGaAs tunnel junctions for multigap cascade solar cells," J. Appl. Phys., vol. 53, pp. 744-748, 1982.

245. G. W. Maden and Ch. E. Backus, "Increased photovoltaic conversion efficiency through use of spectrum splitting and multiple cells," in: Rec. Thirteenth IEEE Photovolt. Spec. Conf., Wash., 1978, pp. 853-857, IEEE, New York, 1978.

246. J. A. Cape, J. S. Harris, and R. Sahai, "Spectrally split tandem converter studies," in: Rec. Thirteenth IEEE Photovolt. Spec. Conf., Wash., 1978, pp. 881-885, IEEE, New York, 1978.

247. Ya. F. Umanskii, X-Ray Micrography of Metals and Semiconductors, Metallurgiya, Moscow, 1969.
248. L. S. Palatnik and V. K. Sorokin, Properties of Semiconducting Thin Films, Energiya, Moscow, 1973.
249. T. M. Golovner, M. B. Kagan, T. L. Lyubashevskaya, T. A. Nuller, and M. N. Kholeva, "Efficiency of p-$Al_xGa_{1-x}As$—n-GaAs heterophotoconverters deduced from optical measurements and photoluminescence spectra," Geliotekhnika [Applied Solar Energy], no. 3, pp. 12-17, 1976.
250. M. K. Antoshin, I. V. Karpenko, O. I. Koval, M. M. Koltun, V. I. Petrov, and M. A. Stepovich, "Scanning electron microscopy of photocells based on cadmium sulfide," Izv. Akad. Nauk SSSR Ser. Fiz., vol. 44, pp. 1290-1293, 1980.
251. M. B. Kagan, A. P. Landsman, and Ya. I. Chernov, "Photoelectric properties of p—n junctions based on diffused single-crystal layers of gallium phosphide," Fiz. Tekh. Poluprovodn., vol. 1, pp. 1335-1341, 1967.
252. A. A. Kukharskii, V. K. Subashiev, and M. B. Ushakova, "A layer of degenerate n-type germanium in heterojunctions," Fiz. Tekh. Poluprovodn., vol. 1, pp. 203-205, 1967.
253. E. Stofel and D. Joslin, "Low-energy proton damage to silicon solar cells," IEEE Trans. Nucl. Sci., vol. 17, pp. 250-255, 1970.
254. G. M. Grigor'eva, L. B. Kreinin, and A. P. Landsman, "Effect of cosmic radiation on solar cells," Geliotekhnika [Applied Solar Energy], no. 5, pp. 3-17, 1971.
255. L. J. Goldhammer and L. W. Slifer, "ATS-6 solar cell flight experiment through 2 years on orbit," in: Rec. 12th IEEE Photovolt. Spec. Conf., Baton Rouge, La., 1976, pp. 199-207, IEEE, New York, 1976.
256. R. L. Crabb and D. Basnett, "Photon induced degradation of electron and proton irradiated silicon solar cells," in: Rec. Tenth IEEE Photovolt. Spec. Conf., Palo Alto, Cal., 1973, pp. 396-403, IEEE, New York, 1974.
257. H. Fisher and W. Pschunder, "Investigation of photon and thermal induced changes in silicon solar cells," in: Rec. Tenth IEEE Photovolt. Spec. Conf., Palo Alto, Cal., 1973, pp. 404-411, IEEE, New York, 1974.
258. W. Pschunder and H. Fischer, "Influence of silicon impurity

content on photon induced avariation of solar cell parameters after particle irradiation," in: Rec. Twelfth IEEE Photovolt. Spec. Conf., Baton Rouge, La., 1976, pp. 270-275, IEEE, New York, 1976.

259. V. G. Weizer, H. W. Brandhorst, J. D. Broder, et al., "Photon degradation effects in terrestrial solar cells," in: Rec. Thirteenth IEEE Photovolt. Spec. Conf., Wash., 1978, pp. 1327-1332, IEEE, New York, 1978.

260. L. J. Cheng, G. B. Turner, R. G. Downing, et al., "Mechanisms of photon-induced changes in silicon solar cell parameters," in: Rec. Thirteenth IEEE Photovolt. Spec. Conf., Wash., 1978, pp. 1333-1336, IEEE, New York, 1978.

261. I. F. Tigane, "Electron-microscope study of conducting SnO_2 layers," Fiz. Tverd. Tela, vol. 7, pp. 276-278, 1965.

262. I. F. Tigane and A. A. Khaav, "Electron-microscope study of the structure of thin sublimated ZnS layers," Izv. Vyssh. Uchebn. Zaved. Fiz., no. 1, pp. 154-155, 1967.

263. K. S. Rebane and I. F. Tigane, "Deposition of thin ZnS films on NaCl cleavage planes in vacuum," Izv. Vyssh. Uchebn. Zaved. Fiz., no. 6, pp. 140-141, 1970.

264. V. M. Efremenkova, I. V. Egorova, and V. E. Yurasova, "Structure and properties of photosensitive CdS layers produced by cathode sputtering," Izv. Akad. Nauk SSSR Ser. Fiz. [Bulletin of the Academy of Sciences of the USSR. Physical Series, Allerton Press, Inc.], vol. 32, pp. 1242-1246, 1968.

265. N. N. Sedov, G. V. Spivak, and V. G. Dyukov, "Measurement of the potential distribution in the p—n junction using the emission electron microscope," Izv. Akad. Nauk SSSR Ser. Fiz. [Bulletin of the Academy of Sciences of the USSR. Physical Series, Allerton Press, Inc.], vol. 32, pp. 1179-1183, 1968.

266. S. P. Shea and L. D. Partain, "Effect of heat treatment on the minority carrier diffusion lengths and junction collection factor in $Cu_x S$/CdS solar cells," Trans. World Elec. Engineering Congress, 21-25 June 1977, Moscow, Section 5A, Informelektro, Moscow, 1977.

267. V. V. Korablev, "Electron spectroscopy of solid surfaces," in: Electronics and Its Applications, VINITI, vol. I2, pp. 3-42, Moscow, 1980 (Itogi Nauki i Tekhniki).

268. K. Siegmahn, C. Nordling, and A. Fahlman, Electron Spectroscopy [Russian translation], I. B. Borovskii (Editor), Mir, Moscow, 1971.

269. M. M. Koltun, Yu. M. Kuznetsov, E. I. Rau, G. V. Sasov, G. V. Spivak, and N. M. Khvastunova, "Scanning electron microscopy of silicon photoconverters," Poverkhnost. Fiz., Khim., Mekh., vol. 1, pp. 70-79, 1982.

270. M. K. Antoshin, V. B. Eliseev, I. V. Karpenko, O. I. Koval, M. M. Koltun, K. N. Puchkova, and M. A. Stepovich, "Scanning electron microscopy of photocells based on cadmium sulfide," in: Eleventh All-Union Conf. on Electron Microscopy, Tallin, 17-19 October 1979, Nauka, vol. 1, Fiz., p. 56, Moscow, 1979.

271. J. J. Loferski, J. Shewchun, E. A. DeMeo, et al., "Characteristics of chalcocite (Cu_xS) films produced by different methods and some properties of solar cells made from such films," in: Rec. Twelfth IEEE Photovolt. Spec. Conf., Baton Rouge, La., 1976, pp. 496-501, IEEE, New York, 1976.

272. N. M. Karelin, S. I. Kusakin, Yu. M. Litvinov, E. I. Rau, and G. V. Spivak, "Electrically active defects in silicon wafers investigated by scanning electron microscopy," Mikroelektronika, vol. 9, pp. 48-53, 1980.

273. S. Martinutsy, F. Kaban-Bruti, T. Kabo, A. Franko, and Zh. Kamiontsis, "Solar cells based on the two-layer CdS—CdZnS system coated with a Cu_2S film," in: Solar Power Engineering [Russian translation], Yu. N. Malevskii and M. M. Koltun (Editors), pp. 332-342, Mir, Moscow, 1979.

274. J. J. Lander, "Auger peaks in the energy spectra of secondary electrons from various materials," Phys. Rev., vol. 91, pp. 1381-1387, 1953.

275. L. A. Harris, "Analysis of materials by electron-excited Auger electrons," J. Appl. Phys., vol. 39, pp. 1419-1427, 1968.

276. A. A. Dorozhkin and N. N. Petrov, "Auger electrons produced under ion implantation of chemical compounds," Poverkhnost. Fiz. Khim. Mekh., vol. 1, pp. 65-70, 1982.

277. N. N. Goryunov and V. G. Grigor'yan, "A television scanning microscope designed to visualize defects in semiconducting

devices," Elektron. Tekh. Ser. 12, Uprav. Kachest. i Stand., no. 3(9), pp. 78-83, 1971.

278. V. G. Grigor'yan, "Scanning optical microscope — a new device for controlling the quality of the surface of semiconductor devices," Elektron. Prom-st., no. 7(I3), pp. 42-44, 1972.

279. M. Martin and H. Williams, "Optical scanning of silicon wafer for surface contaminations," Electron-Opt. Syst. Design, Sept., pp. 45-49, 1980.

280. A. S. Maksimov and A. B. Ormont, "A study of inversion regions on the surface of silicon p^+—n structures using a scanning optical microscope," Mikroelektronika, vol. 9, pp. 155-159, 1980.

281. J. F. Allison, R. A. Arndt, and H. A. Meulenberg, "Comparison of the Comsat violet and nonreflective cells," in: Rev. Tenth IEEE Photovolt. Spec. Conf., Palo Alto, Cal., 1973, pp. 1038-1041, IEEE, New York, 1974.

282. N. M. Bordina, A. K. Zaitseva, E. A. Marasanova, and A. A. Polisan, "Silicon photoconverters with a texturized surface and their properties," Geliotekhnika [Applied Solar Energy], no. 3, pp. 6-11, 1982.

283. D. Kh. Morosov, T. Ya. Ryabova, and V. V. Tsetlin, "Some aspects of active shielding against the radiation in space," Sci. Rep. CERN, no. 16/1, pp. 501-507, 1971.

284. A. A. Abdullin, G. I. Artamonova, M. M. Koltun, A. I. Lezikhin, T. K. Pavlushina, V. I. Red'ko, V. V. Tsetlin, and M. S. Emishyan, "Protective properties of dielectric materials with a space charge," in: Third All-Union Conf. on the Radiological Protection in Nuclear Engineering, Tbilisi, 27-29 October 1981: Abstracts, Institute of Applied Mathematics, Tbilisi State University, p. 74, 1981.

285. V. V. Tsetlin, "Back reflection of fast electrons from high-resistivity dielectrics," in: Third All-Union Scientific Conf. on Radiological Protection in Nuclear Engineering, Tbilisi, 27-29 October 1981 (Abstracts), Institute of Applied Mathematics, p. 104, Tbilisi State University, 1981.

286. L. Holland, Vacuum Deposition of Thin Films [Russian translation], N. V. Vasil'chenko (Editor), Gosenergoizdat, Moscow, 1963.

287. N. V. Suikovskaya, Chemical Methods of Producing Thin Transparent Films, Khimiya, Leningrad, 1971.

288. N. Mardesich, "Solar cell efficiency enhancement by junction etching and conductive AR coating processes," in: Rec. Fifteenth IEEE Photovolt. Spec. Conf., Kissimee, Fla., 1981, pp. 446-449, IEEE, New York, 1981.

289. M. M. Koltun, V. P. Matveev, and E. A. Agaev, "Two-layer antireflective silicon photocells," Geliotekhnika [Applied Solar Energy], no. 5, pp. 36-38, 1982.

290. M. M. Koltun and T. M. Golovner, "Antireflective silicon photocells," Opt. Spektrosk., vol. 21, pp. 630-637, 1966.

291. J. T. Cox and G. M. Hass, "Antireflective coatings for optical and infrared optical materials," in: Physics of Thin Films, G. M. Hass and R. E. Thun (Editors), Academic Press, 1964.

292. I. V. Grebenshchikov, A. G. Vlasov, B. S. Neporent, and N. V. Suikovskaya, Antireflective Optics, Gostekhizdat, Moscow-Leningrad, 1946.

293. O. S. Heavens, Optical Properties of Thin Solid Films, Butterworth, 1955.

294. G. V. Rozenberg, Optics of Thin-Layer Coatings, Fizmatgiz, Moscow, 1958.

295. P. H. Berning, "Theory and calculations of optical thin films," in: Physics of Thin Films, G. M. Hass (Editor) [Russian translation], pp. 51-151, Mir, Moscow, 1967.

296. M. M. Koltun, "Multilayer antireflective coatings for the receiving surface of semiconductor photocells," Geliotekhnika [Applied Solar Energy], no. 4, pp. 14-17, 1969.

297. R. N. Tykvenko, M. M. Koltun, V. P. Matveev, and L. S. Serkh, "Antireflective thin films for semiconductor thin-film photocells and the protection of p—n junctions during the deposition of contacts," in: Thin Films and Their Applications: Proc. Second Republican Conf., Vilnius, 2 December 1968, Academy of Sciences of the Lithuanian SSR, pp. 201-202, Vilnius, 1969.

298. J. Lindmayer and C. A. Wrigley, "A new lightweight solar cell," in: Rec. Twelfth IEEE Photovolt. Spec. Conf., Baton Rouge, La., 1976, pp. 53-54, IEEE, New York, 1976.

299. M. M. Koltun and I. P. Gavrilova, "Antireflective coatings for metallized surface of solar cells and thermal collectors,"

Geliotekhnika [Applied Solar Energy], no. 4, pp. 44-48, 1982.

300. M. M. Koltun and I. P. Gavrilov, "Determination of the parameters of antireflective coatings for solar cells with heterostructures," Zh. Prikl. Spektrosk., vol. 41, 1984.

301. N. S. Lidorenko, S. V. Ryabikov, G. S. Daletskii, A. I. Kozlov, M. M. Koltun, V. P. Matveev, and M. V. Nikiforova, "Optimization of the optical and thermophysical parameters of coatings for electromagnetically clean solar batteries," Geliotekhnika, [Applied Solar Energy], no. 1, pp. 3-5, 1983.

302. M. Wolf, "Solar energy residential system modeling," in: Proc. Intern. Congr. "The Sun in the service of mankind," The Section "The Photovoltaic power and its application in space and on the Earth," Cent. Nat. Etudes Spatiales, pp. 463-476, Bretigny-sur-Orge, 1973.

303. K. W. Boer, "Direct solar energy conversion for terrestrial use," J. Environ. Sci., vol. 17, pp. 8-14, 1974.

304. M. M. Koltun and I. P. Gavrilova, "Optimization of electrical and optical parameters of silicon photocells for photothermal converters of concentrated solar radiation," Geliotekhnika [Applied Solar Energy], no. 1, pp. 3-12, 1978.

305. M. M. Koltun, I. P. Gavrilova, and M. Kolenkin, "Silicon photocells with selective multilayer coatings," Zh. Prikl. Spektrosk., vol. 37, pp. 340-343, 1982.

306. M. M. Koltun, V. P. Matveev, and I. P. Gavrilova, "Photothermal collectors of solar radiation," Geliotekhnika [Applied Solar Energy], no. 5, pp. 3-11, 1980.

307. M. M. Koltun, U. Kh. Gaziev, and Sh. A. Faiziev, "Glass and polymer insulation of solar power installations," Geliotekhnika [Applied Solar Energy], no. 1, pp. 42-48, 1975.

308. M. M. Koltun and Sh. A. Faiziev, "Optical heat-reflecting coatings deposited by evaporation in a vacuum," Opt.-Mekh. Prom-st., no. 7, pp. 39-41, 1975.

309. M. D. Kudryashova, "New selective coatings for collector surfaces in solar power installations," Geliotekhnika, no. 4, pp. 47-56, 1969.

310. M. M. Koltun, "Black multilayer mirror," Zh. Prikl. Spektrosk., vol. 12, pp. 350-352, 1970.

311. G. A. Gukhman and M. M. Koltun, "Selective coatings for

thermal solar energy converters," Geliotekhnika, no. 4, pp. 3-5, 1983.

312. B. O. Seraphin, "Selective optical surfaces and their role in photothermal conversion of solar radiation," in: Solar Energy Converters (Problems in Solid-State Physics) [Russian translation], M. M. Koltun and V. M. Evdokimov (Editors), Energoizdat, pp. 8-55, Moscow, 1982.

313. M. M. Koltun, L. A. Ryabova, and E. A. Agaev, "Selective optical coatings based on black cobalt oxide films produced by pyrolysis," Geliotekhnika [Applied Solar Energy], no. 6, pp. 28-30, 1982.

314. C. Choudbury and H. K. Sehgal, "Black cobalt selective coatings by spray pyrolysis for photothermal conversion of solar energy," Solar Energy, vol. 28, pp. 25-31, 1982.

315. M. D. Kudryashova, "Mechanical treatment of collector surfaces for solar energy conversion systems, designed to improve the selectivity of optical properties," Geliotekhnika, no. 5, pp. 36-39, 1969.

316. L. R. Chapman and G. L. Vaneman, "Solar collector based on whisker-shaped oxides, grown on metallic substrates," Solar Energy, vol. 28, pp. 77-79, 1982.

317. L. S. Palatnik, I. Kh. Tartakovskaya, and O. I. Kovaleva, "Stable selective coatings based on high vacuum deposition of metals," in: Solar Energy Conversion: Abstracts of Papers read at the Chernogolovka Conf. on 17-19 February 1981, Institute of Chemical Physics, USSR Academy of Sciences, pp. 213-214, Chernogolovka, 1981.

318. L. S. Palatnik, I. Kh. Tartakovskaya, O. I. Kovaleva, P. G. Cheremskoi, and A. S. Derevyanchenko, "Structural properties of highly absorbing film materials based on aluminum and deposited in vacuum," in: Solar Energy Conversion: Abstracts of Papers read at the Chernogolovka Conf. on 17-19 February 1981, Institute of Chemical Physics, USSR Academy of Sciences, pp. 214-215, Chernogolovka, 1981.

319. L. Melamed, "Survey of selective absorber coatings for solar energy technology," J. Energ., vol. 1, pp. 100-107, 1977.

320. Soviet Patent No. 868282 (USSR), Multilayer Selective Coatings for Solar Collectors, M. M. Koltun, G. A. Gukhman, Yu. N. Malevskii, N. G. Milevskaya, M. D. Kudryashova,

V. S. Sinyavskii, K. I. Makarova, and G. T. Eidinova, Submitted 13 December 1979, No. 2852908, publ. in Byull. Izobret., no. 36, p. 145, 1981.

321. M. M. Koltun, G. A. Gukhman, V. S. Sinyavskii, N. G. Milevskaya, G. T. Eidinova, and K. I. Makarova, "Selective coatings for heat-receiving surfaces of flat solar collectors made from the AD1 alloy," Tekh. Leg. Splavov, no. 1, pp. 58-61, 1982.

322. G. A. Gukhman, M. M. Koltun, A. I. Malik, and M. I. Umarova, "Two-layer coating for thermal solar collectors, produced by hydrolysis," Geliotekhnika [Applied Solar Energy], no. 5, pp. 37-38, 1983.

323. M. M. Koltun and Sh. A. Faiziev, "Utilization of transparent heat-reflecting coatings in solar energy conversion," Geliotekhnika [Applied Solar Energy], no. 1, pp. 28-31, 1977.

324. M. M. Koltun and I. P. Gavrilova, "Optimization of the optical properties of multilayer selective coatings," Zh. Prikl. Spektrosk., vol. 34, pp. 749-751, 1981.

325. M. M. Koltun and I. P. Gavrilova, "Optical parameters of selective coatings for thermal solar collectors," Geliotekhnika [Applied Solar Energy], no. 3, pp. 35-39, 1982.

326. I. N. Shklyarevskii and V. G. Padalka, "Measurement of the optical constants of copper, gold, and nickel in the infrared," Opt. Spektrosk., vol. 6, pp. 78-83, 1959.

327. S. Roberts, "Optical properties of nickel and tungsten and their interpretation according to Drude's formula," Phys. Rev., vol. 114, pp. 104-115, 1959.

328. E. M. Lushiku and K. R. O'Shea, "Ellipsometry in the study of selective radiation-absorbing surfaces," Solar Energy, vol. 19, pp. 271-276, 1977.

329. M. M. Koltun, V. P. Molchanova, F. R. Yuppets, and I. P. Gavrilova, "Parameters of electrochemical coatings for solar collectors," Geliotekhnika [Applied Solar Energy], no. 6, pp. 84-85, 1979.

330. M. Okuyama, K. Furusawa, and Y. Hamakawa, "Ni cermet selective absorbers for solar photothermal conversion," Solar Energy, vol. 22, pp. 479-482, 1979.

331. M. M. Koltun and I. P. Gavrilova, "Effect of spectral composition of solar radiation on the parameters of optically selec-

tive absorbing surfaces," Zh. Prikl. Spektrosk., vol. 41, pp. 849-852, 1984.

332. M. P. Thekaekara, "Extraterrestrial solar irradiance," in: Solar Cells, p. 2, IEEE, New York, 1976.

333. H. Brandhorst, J. Hickey, H. Curtis, and E. Ralph, "Interim solar cell testing procedures for terrestrial applications," in: ERDA/NASA Workshop on Terrestrial Photovoltaic Measurements, pp. 1-15, Cleveland, Ohio, 1975.

334. M. Van der Leij and C. J. Hoogendoorm, "Influence of the direct spectral solar distribution on the normal total absorptivity of spectral selective surfaces," Solar Cells, vol. 19, pp. 575-577, 1977.

335. S. A. Demidov, B. A. Khrustalev, and N. B. Rekant, "A simple portable device for measuring the radiative power of solids at room temperature," Geliotekhnika [Applied Solar Energy], no. 6, pp. 36-43, 1971.

336. K. O. Bartsch and W. Hudgins, "Investigation of the emittance of coated refractory metals," AIAA Paper N 70-68, pp. 1-10, 1970.

337. V. A. Ospova, Experimental Study of Heat Transfer, Energiya, Moscow-Leningrad, 1969.

338. M. M. Koltun and E. A. Agaev, "Radiative power of materials and selective coatings for solar power engineering at working temperatures," Geliotekhnika [Applied Solar Energy], no. 1, pp. 17-19, 1984.

339. C. N. Watson-Munro and C. M. Horwitz, "Selective surfaces," in: Solar Energy, pp. 291-313, Pergamon Press, Oxford, 1975.

340. V. A. Baum and M. B. Bektenev, "Effect of selective coatings on the heat balance of solar collectors," Izv. Akad. Nauk Turkm. SSR Ser. Fiz.-Tekh. Khim. Geol. Nauk, no. 1, pp. 21-26, 1970.

341. M. M. Koltun, O. A. Nevezhin, A. V. Romankevich, and E. M. Yurin, "Increasing the efficiency of conversion of solar radiation into thermal power using evacuated glass tubular cells," Geliotekhnika [Applied Solar Energy], no. 4, pp. 3-4, 1980.

342. Soviet Patent No. 851012 (USSR), Composite Solar Collector, M. M. Koltun, V. P. Matveev, and I. P. Gavrilova, sub-

mitted 5 November 1979, No. 2839599, publ. in Byull. Izobret., no. 28, p. 150, 1981.

343. I. P. Gavrilova, M. M. Koltun, and V. P. Matveev, "High-efficiency composite solar converters," in: Solar Energy Conversion: Abstracts of Papers read at the Chernogolovka Conf. on 17-19 February 1981, Institute of Chemical Physics, USSR Academy of Sciences, p. 60, Chernogolovka, 1981.

344. D. C. Carmichael, G. B. Gaines, F. A. Sliemers, and C. W. Kistler, "Materials for encapsulation systems for terrestrial photovoltaic arrays," in: Rec. Twelfth IEEE Phtovolt. Spec. Conf., Baton Rouge, La., 1976, pp. 317-323, IEEE, New York, 1976.

345. G. V. Schumann and R. Bus, "A program for perfecting the fabrication technology for terrestrial solar electric generators in the Federal Republic of Germany," in: Solar Power Engineering [Russian translation], Yu. N. Malevskii and M. M. Koltun (Editors), pp. 380-387, Mir, Moscow, 1979.

346. N. V. Pul'manov and V. N. Potapov, "Solar batteries in protective envelopes," Geliotekhnika [Applied Solar Energy], no. 5, pp. 25-28, 1971.

347. G. S. Daletskii, I. V. Karpenko, and M. M. Koltun, "Degradation of the electrophysical parameters of photoconverters in the course of prolonged utilization," Geliotekhnika [Applied Solar Energy], no. 5, pp. 7-12, 1979.

348. The Pravda, 8 August 1981.

349. The Pravda, 22 September 1981.

350. N. S. Lidorenko, S. V. Ryabikov, G. S. Daletsky, et al., "Optimization of optical and thermophysical coating characteristics for 'electro-magneto-clean' solar batteries," in: Proc. Condensed Paper. Fifth Miami Intern. Conf. on Alternative Energy Sources, Miami Beach, Fla., 1982, Clean Energy Res. Inst., p. 105, Coral Gables, Fla., 1982.

351. Yu. Zhuk, "Biological laboratory in orbit," The Izvestiya, 4 August 1977.

352. M. M. Koltun and V. V. Tsetlin, "Stability of the charged state of optical coatings for solar cells in space," Geliotekhnika [Applied Solar Energy], no. 2, 1985.

353. K. A. Dergabuzov, O. B. Evdokimov, and B. A. Kononov,

Radiation Diagnostics of Electric Potentials, Atomizdat, Moscow, 1978.

354. G. S. Daletskii, V. A. Karpukhin, M. M. Koltun, and V. M. Kuznetsov, "Experimental study of the effects of solar radiation reflected from the Earth and of its cloud cover on the heat balance of solar cells on one of the Cosmos satellites," Geliotekhnika [Applied Solar Energy], no. 5, pp. 3-6, 1982.

355. Astronomical Calendar. Permanent Part, Nauka, Moscow, 1973.

356. M. P. Thekaekara and A. J. Drummond, "Standard values for the solar constant and its spectral components," Nature. Phys. Sci., vol. 229, pp. 483-492, 1971.

357. M. P. Thekaekara, "Extraterrestrial solar spectrum, 3000-6100 Å at 1 Å interval," Appl. Opt., vol. 13, pp. 518-522, 1974.

358. E. A. Makarova and A. V. Kharitonov, Distribution of Energy in the Solar Spectrum and the Solar Constant, Nauka, Moscow, 1972.

359. C. W. Allen, Astrophysical Quantities [Russian translation], Mir, Moscow, 1972.

360. A Model of Solar Radiation (for Power Calculations), Vavilov State Optical Institute, Leningrad, 1979.

361. E. A. Makarova and A. V. Kharitonov, "Comparison of present-day average parameters of solar radiation," Astron. Zh., vol. 52, pp. 965-969, 1975.

362. K. Bogus, J. C. Larue, and R. L. Crabb, "Solar cell calibration: recent experiences at ESTEC and proposal of combined space and terrestrial calibration procedure," in: Proc. Intern. Photovolt. Solar Energy Conf., Luxembourg, 27-30 September 1977, pp. 754-768, Riedel Pub. Co., Dordrecht-Boston, 1978.

363. J. R. Hickey, "A review of solar constant measurement: Sun, mankind's future source of energy," in: Proc. Intern. Solar Energy Soc. Congr., New Delhi, India, January 1978, vol. 1, pp. 331-337, Pergamon Press, New York, 1978.

364. M. P. Thekaekara, "Solar irradiance: total and spectral and its possible variations," Appl. Opt., vol. 15, pp. 915-920, 1976.

365. O. V. Vasil'ev, A. F. Muragin, G. A. Nikol'skii, and B. M.

Rubashev, "Possible variations in the solar constant," Solnech. Dannye, no. 3, pp. 80-85, 1973.

366. M. M. Koltun and I. S. Orshanskii, "Metrology of solar cells," Geliotekhnika [Applied Solar Energy], no. 3, pp. 3-13, 1981.

367. M. G. Kroshkin, Physicotechnical Principles for Space Studies, Mashinostroenie, Moscow, 1969.

368. L. V. Kozlov, M. D. Nusinov, A. I. Akishin, V. M. Zaletaeva, V. V. Kozelkin, and E. N. Evlanov, Simulation of Thermal Conditions on a Spacecraft and in the Ambient Medium, G. I. Petrov (Editor), Mashinostroenie, Moscow, 1971.

369. F. S. Johnson, "The solar constant," J. Meteorol., vol. 11, pp. 431-435, 1954.

370. K. Bogus, Solar Constant, AM0 Spectral Irradiance and Solar Cell Calibration, ESA, Noordwijk, The Netherlands, 1975.

371. Yu. D. Yanishevskii, Actinometric Instrumentation and Methods, Gidrometeoizdat, Leningrad, 1957.

372. R. Penndorf, "Tables of the refractive index for standard air and the Rayleigh scattering coefficient for the spectral region between 0.2 and 20 μm and their application at atmospheric optics," J. Opt. Soc. Am., vol. 47, pp. 176-181, 1957.

373. Ch. P. Brichambaut, Solar Radiation and Radiation Transfer in the Atmosphere [Russian translation], M. S. Malkevich (Editor), Mir, Moscow, 1966.

374. Bo Leckner, "The spectral distribution of solar radiation at the Earth's surface — elements of a model," Solar Energy, vol. 20, pp. 143-150, 1978.

375. A. Angstrom, "On the atmospheric transmission of sun radiation and on the dust in the air," Geogr. Ann., vol. 11, pp. 156-164, 1929.

376. M. P. Thekaekara, "Solar radiation measurement: techniques and instrumentation," Solar Energy, vol. 18, pp. 309-325, 1976.

377. D. M. Gates and W. J. Harrop, "Infrared transmission of the atmosphere to solar radiation," Appl. Opt., vol. 2, pp. 887-888, 1963.

378. D. M. Gates, "Near infrared atmospheric transmission to solar radiation" J. Opt. Soc. Am., vol. 50, pp. 1299-1304, 1960.

379. R. Bird and R. Hulstrom, Direct Insolation Models, Solar Energy Res. Inst., SERI/TR-335-344, Golden, Col., 1980.

380. P. Moon, "Proposed standard solar-radiation for engineering use," J. Franklin Inst., vol. 230, pp. 583-587, 1940.

381. K. W. Boer, "The solar spectrum on typical clear weather days," Solar Energy, vol. 19, pp. 525-538, 1977.

382. C. J. Kok, "Spectral irradiance of daylight for air mass 2," J. Phys. D: Appl. Phys., vol. 5, p. L85, 1972.

383. C. J. Kok, "Spectral irradiance of daylight at Pretoria," J. Phys. D: Appl. Phys., vol. 5, pp. 1513-1520, 1972.

384. "Recommendations concernant l'eclairement energetique et la repartition spectrale du rayonnement solaire en vue de sa reproduction artificielle pour des essais," Bubl. CIE, no. 20, pp. 1-54, 1972.

385. D. N. Lazarev, "Characteristics of solar radiation," Svetotekhnika, no. 8, pp. 8-11, 1976.

386. J. Krochmann and M. Seidl, "Quantitative data on daylight for illuminating engineering," Light. Res. Technol., vol. 6, pp. 165-171, 1974.

387. H. Hadley, "Proposed standard for AM1 sunlight," in: Terrestrial Photovoltaic Measurements Workshop Proc., 19-21 March 1975, Cleveland, Ohio, NASA TM-71802, pp. 80-85, 1975.

388. D. M. Gates, "Spectral distribution of solar radiation at the Earth's surface," Science, vol. 151, pp. 523-527, 1966.

389. H. Brandhorst, J. Hickey, H. Curtis, and E. Ralph, "Interim solar cell testing procedures for terrestrial applications," Technical Memorandum 71771, NASA, Cleveland, Ohio, 1975.

390. H. W. Brandhorst, "Terrestrial solar cell calibration and measurement procedures," in: Proc. Intern. Photovolt. Solar Energy Conf., Luxemburg, 27-30 September 1977, pp. 745-753, Reidel Publ. Co., Dordrecht-Boston, 1978.

391. Proc. International Electrotechnical Commission, Technical Committee No. 82 on Solar Photogalvanic Power Systems, Geneva, IEC, 1982. See also "Terrestrial photovoltaic measurement procedures," Technical Memorandum 73702, NASA, Cleveland, Ohio, 1977.

392. F. C. Treble, "Terrestrial photovoltaic performance mea-

surement," in: Proc. Intern. Photovolt. Solar Energy Conf., Luxemburg, 27-30 September 1977, pp. 732-744, Riedel Publ. Co., Dordrecht; Boton, 1978.

393. M. M. Koltun and I. S. Orshanskii, "International standardization of measurements on solar cells for terrestrial applications," Geliotekhnika [Applied Solar Energy], no. 2, pp. 83-86, 1981.

394. V. Ya. Koval'skii, "Solar simulators and measurement of the parameters of solar batteries and their elements (Review)," Geliotekhnika [Applied Solar Energy], no. 3, pp. 45-51, 1972.

395. V. Ya. Koval'skii and D. A. Shklover, "AM0 solar simulator," Geliotekhnika [Applied Solar Enery], no. 1, pp. 35-42, 1967.

396. V. Ya. Koval'skii, "Precision of simulation of solar radiation," in: All-Union Conf. on Solar Energy Utilization, Erevan, 17-21 June 1969, Sec. 6, pp. 61-67, VNIIT, Moscow, 1969.

397. N. A. Valyus, Scanning Optical Instruments, Mashinostroenie, Moscow, 1966.

398. "Spectrosun model X-25 solar simulator," Data Sheet 1103A, Spectrolab., Silmar, Cal., 1964.

399. D. W. Ritchie and J. D. Sandstrom, "On evaluation of photovoltaic devices for future spacecraft power demands," Energy Convers., vol. 9, pp. 83-90, 1969.

400. N. I. Zhigalina and A. I. Rymov, "Solar simulator," Elektrotekh. Prom-st. Ser. Svetotekh. Izdel., no. 5(41), pp. 2-3, 1976.

401. S. N. Molodtsov, L. B. Prikupets, and Yu. A. Bukhanov, "Solar simulator for the visible and near-infrared," Elektrotekh. Prom-st. Ser. Svetotekh. Izdel., no. 1(37), pp. 4-6, 1976.

402. G. A. Naraikina, "Metallogenic lamps with a continuous emission spectrum," Elektrotekh. Prom-st. Ser. Svetotekh. Izdel., no. 5(59), pp. 11-22, 1979.

403. Soviet Patent No. 727939 (USSR), An Illuminator, V. S. Luk'yanenko, A. I. Ivanov, N. S. Orshanskii, E. V. Kraev, S. M. Eroshin, A. I. Antonov, T. A. Kozyreva, and Yu. N. Baranov, submitted 20 May 1976, No. 2363527, publ. in Byull. Izobret., no. 14, p. 187, 1980.

404. Soviet Patent 679907 (USSR), Optical Filter, V. S. Luk'ya-nenko, A. I. Ivanov, E. V. Kraev, S. M. Eroshin, A. I. Antonov, V. M. Dolgov, I. S. Orshanskii, and Yu. N. Baranov, submitted 15 January 1976, no. 2328409, publ. in Byull. Izobret., no. 30, p. 168, 1979.

405. A. S. Ivantsev, "Application of selective coatings in light sources," Svetotekhnika, no. 12, pp. 1-5, 1979.

406. L. N. Aleksandrov and A. S. Ivantsev, Multilayer Thin-Film Structures for Light Sources, Nauka, Novosibirsk, 1981.

407. T. I. Buyankina, A. S. Ivantsev, and V. I. Konyashkina, "A lamp simulator of solar radiation," Svetotekhnika, no. 6, pp. 7-8, 1977.

408. H. Curtis, "Low cost AM2 simulator," in: Terrestrial Photovoltaic Measurements Workshop Proc., 19-21 March 1975, pp. 98-103, NASA TM-71802, Cleveland, Ohio, 1975.

409. R. W. Opiorden, "Large area pulsed solar simulator," in: Rec. Eighth IEEE Photovolt. Spec. Conf., Seattle, Wash., 1970, pp. 312-318, IEEE, New York, 1970.

410. Soviet Patent No. 509842 (USSR), A Device for Testing Semiconductor Solar Batteries, B. A. Kryzhanovskii, Yu. P. Dorofeev, and V. Ya. Koval'skii, submitted 16 July 1974, no. 2046482, publ. in Byull. Izobret., no. 7, p. 175, 1976.

411. Yu. M. Belyaev, Yu. P. Dorofeev, A. P. Gura, and B. A. Kryzhanovskii, "Pulsed solar simulator for testing solar batteries," in: Pulsed Photometry, no. 6, pp. 249-252, Mashinostroenie, Leningrad, 1979.

412. A. D. Haigh and J. M. Shaw, "A low cost solar simulator for testing photovoltaic terrestrial power cells and modules," in: Proc. Second E. C. Photovolt. Solar Energy Intern. Conf., West Berlin, 23-26 April 1979, pp. 487-494, Reidel Publ. Co., Dordrecht, 1979.

413. G. I. Rabinovich, "A 40-50-kW high-pressure xenon lamp," Svetotekhnika, no. 9, pp. 10-13, 1969.

414. D. Bickler, "The simulation of solar radiation," Solar Energy, vol. 4, pp. 64-68, 1962.

415. H. Hadley, "Spectral effects in CdS/Cu_2S solar cells," in: Terrestrial Photovoltaic Measurements Workshop Proc., 19-21 March 1975, pp. 113-119, NASA TM 71802, Cleveland, Ohio, 1975.

416. J. A. McMillan and E. M. Peterson, "Solar simulation with tungsten-halogen quartz lamps and optical filters," Solar Energy, vol. 22, pp. 467-469, 1979.

417. A. Seck, "Spectral distribution of sunlight under various air mass conditions," in: Terrestrial Photovoltaic Measurements Workshop Proc., 19-21 March 1975, pp. 25-27, NASA TM-71802, Cleveland, Ohio, 1975.

418. T. A. Kozyreva, M. M. Koltun, and I. S. Orshanskii, "Calibration of standards for terrestrial measurements of solar cell parameters," Geliotekhnika [Applied Solar Energy], no. 1, pp. 12-14, 1980.

419. I. V. Gracheva, A. I. Ivanov, M. M. Koltun, G. D. Naumova, and I. S. Orshanskii, "Selective radiometers for the adjustment of solar simulators," Svetotekhnika, no. 4, pp. 11-12, 1981.

420. H. B. Curtis, "Global calibration of terrestrial reference cells and errors involved in using different irradiance monitoring techniques," in: Rec. Fourteenth IEEE Photovolt. Spec. Conf., San Diego, Cal., 1980, pp. 500-505, IEEE, New York, 1980.

421. "Standard silicon solar cell by International Rectifier Corporation," Sun at Work, No. 4, pp. 15-18, 1966.

422. Soviet Patent No. 800679 (USSR), A. I. Ivanov, G. V. Vasil'ev, M. M. Koltun, I. S. Orshanskii, and G. D. Naumova, submitted 3 January 1979, no. 2705572, publ. in Byull. Izobret., no. 4, p. 152, 1981.

423. M. M. Koltun, V. P. Matveev, and I. S. Orshanskii, "Measurement of the parameters of silicon solar cells in Socialist countries," Geliotekhnika [Applied Solar Energy], no. 3, pp. 3-5, 1983.

424. D. L. Gendelev, I. V. Gracheva, M. M. Koltun, I. S. Orshanskii, and L. B. Serebrova, "A selective radiometer for field tests of terrestrial solar batteries," in: Fourth All-Union Scientific-Technological Conf. on Photometry and Its Metrological Applications, Moscow, 29 November—3 December 1982. Abstracts, p. 385, VNIIOFI, Moscow, 1982.

425. I. V. Grachev, M. M. Koltun, and I. S. Orshanskii, "Stability of the parameters of selective radiometers with photoelectric detectors," Izmer. Tekh. Ser. Metrologiya, no. 1, 1985.

426. N. L. Thomas and F. W. Sarles, "High altitude calibration of thirty-three silicon and gallium arsenide solar cells on a sounding rocket," in: Rec. Twelfth IEEE Photovolt. Spec. Conf., Baton Rouge, La., 1976, pp. 560-568, IEEE, New York, 1976.

427. F. W. Sarles, W. C. Haase, and P. F. McKenzie, "Balloon flight instrumentation for solar cell measurements," Rev. Sci. Instrum., vol. 42, pp. 346-351, 1971.

428. H. W. Brandhorst, "Solar cell calibrated on high altitude aircraft," Space Aeronaut., vol. 45, pp. 122-123, 1966.

429. K. H. David, "Solar cell calibration for AM0 short-circuit in terrestrial sunlight," Sci. Industr. Spat., vol. 4, pp. 31-41, 1968.

430. V. Ya. Koval'skii, "Methods of calibrating irradiance standards," Geliotekhnika [Applied Solar Energy], no. 6, pp. 66-71, 1969.

431. E. A. Makarova, N. I. Kozhevnikov, and G. A. Porfir'eva, "Determination of AM0 output power of silicon photocells," Soobshch. Shternberg State Astronomical Institute, no. 116, p. 25-45, 1961.

432. V. Ya. Koval'skii, E. V. Kononovich, I. S. Orshanskii, and N. N. Shakura, "A method of taking into account the Forbes effect in the determination of the AM0 values of the current output of silicon photocells," Soobshch. Shternberg State Astronomical Institute, no. 214-215, pp. 3-12, 1979.

433. K. Ya. Kondrat'ev, Radiant Solar Energy, Gidrometeoizdat, Leningrad, 1954.

434. E. V. Pyaskovskaya-Fesenkova, Scattering of Light in the Earth's Atmosphere, USSR Academy of Sciences, Moscow, 1957.

435. Astronomical Calendar 1975 (Variable Part), Nauka, Moscow, 1974.

436. F. C. Treble, "Optical aspects of solar cells performance measurement," Sci. Industr. Spat., vol. 1, pp. 37-47, 1965.

437. W. Arndt, W. H. Bloss, and G. H. Hewig, "Determination of the spectral distribution of global radiation with a rapid spectral radiometer and its correlation with solar cell efficiency," in: Proc. Second E. C. Photovolt. Solar Intern. Conf.,

West Berlin, 23-26 April 1979, pp. 987-994, Reidel Publ. Co., Dordrecht, 1979.

438. B. Umarov, M. Kagan, M. Koltun, and M. Umarova, "High mountain of solar cells," in: Proc. Condensed Pap. Fourth Miami Intern. Conf. on Alternative Energy Sources, 14-16 December 1981, Miami Beach, Fla., Clean Energy Res. Inst., pp. 214-216, Coral Gables, Fla., 1981.

439. M. M. Koltun and I. S. Orshanskii, "Metrological quality of measurements of the photoelectric parameters of solar cells for terrestrial applications," in: Fourth All-Union Scientific-Technological Conf. on "Photometry and Its Metrological Applications," Moscow, 29 November—3 December 1982. Abstracts, p. 324, VNIIOFI, Moscow, 1982.

440. H. K. Gummel and F. M. Smits, "Evaluation of solar cells by means of spectral analysis," Bell Syst. Tech. J., vol. 43, pp. 1103-1107, 1964.

441. H. W. Brandhorst, "Prediction of terrestrial solar cell short-circuit currents by spectral analysis," in: Terrestrial Photovoltaic Measurements Workshop Proc., 19-21 March 1975, pp. 120-128, NASA TM-71802, Cleveland, Ohio, 1975.

442. R. E. Hart, "Solar cell measurement techniques used at NASA Lewis Research Center," in: Terrestrial Photovoltaic Measurements Workshop Proc., 19-21 March 1975, pp. 134-139, NASA TM-71802, Cleveland, Ohio, 1975.

443. J. C. Larue, "Pulsed measurement of solar cell spectral response," in: Proc. Second E. C. Photovolt. Solar Energy Intern. Conf., West Berlin, 23-26 April 1979, pp. 477-486, Reidel Publ. Co., Dordrecht, 1979.

444. V. L. Daval and A. Rothwarf, "Comment on simple measurement of absolute solar cell efficiency," J. Appl. Phys., vol. 50, pp. 1822-1823, 1979.

445. E. B. Vinogradova, T. M. Golovner, S. M. Gorodetskii, and L. B. Kreinin, "Photoelectric parameters of photoconverters as functions of irradiance," Geliotekhnika [Applied Solar Energy], no. 2, pp. 13-17, 1979.

446. E. B. Vinogradova, T. M. Golovner, S. M. Gorodetskii, and N. S. Zhdanovich, "Determination of the recombination parameters of a photocell under different injection levels," Prib. Tekh. Eksp., no. 6, pp. 153-154, 1976.

447. E. F. Zalewsky and J. Geist, "Solar cell response characterization," Appl. Opt., vol. 18, pp. 3942-3947, 1979.

448. T. M. Golovner, M. M. Koltun, A. L. Kostanenko, and I. S. Orshanskii, "Determination of the integrated current from solar cells using improved measurements of spectral sensitivity," Zh. Prikl. Spektrosk., vol. 37, pp. 471-475, 1982.

449. B. P. Kozyrev, "Fundamentals of the calculation and fabrication of a radiation thermocouple," Izv. LETI, no. 44, pp. 124-129, 1960.

450. I. M. Vesel'nitskii, Yu. D. Ignat'ev, A. S. Il'in, S. M. Mel'nikova, and A. B. Fromberg, "The PP-2 measuring primary strip converter," in: Third-All Union Scientific-Technological Conf. on Photometry and Its Metrological Applications, Moscow, 2-6 December 1979: Abstracts, p. 168, VNIIOFI, Moscow, 1979.

451. M. V. Egorova, L. N. Samoilov, L. E. Svyatova, B. M. Stepanov, R. I. Stolyarevskaya, V. P. Frolov, and M. I. Epshtein, "High-precision systems for testing the measurements of relative spectral parameters of sources and radiation detectors," in: Third All-Union Scientific-Technological Conf. on Photometry and Its Metrological Applications, Moscow, 2-6 December 1979: Abstracts, p. 168, VNIIOFI, Moscow, 1979.

452. E. G. Polyakova, L. E. Svyatova, R. I. Stolyarevskaya, and V. P. Frolov, "A system for measuring the sensitivity and linearity of optical radiation detectors," in: Third All-Union Scientific-Technological Conf. on Photometry and Its Applications, Moscow, 2-6 December 1979: Abstracts, p. 157, VNIIOFI, Moscow, 1979.

453. O. A. Minaeva, L. N. Samoilov, L. E. Svyatova, B. M. Stepanov, V. S. Fel'dman, V. P. Frolov, and M. I. Epshtein, "A reference source of UV radiation at 0.254 μm," in: Third All-Union Scientific-Technological Conf. on Photometry and Its Applications, Moscow, 2-6 December 1979: Abstracts, p. 100, VNIIOFI, Moscow, 1979.

454. O. A. Minaeva, L. E. Svyatova, and V. P. Frolov, "Spectral sensitivity of type RTN thermal detectors in the ultraviolet visible and near-infrared regions," Fourth All-Union Scientific-Technological Conf. on Photometry and Its Applications,

Moscow, 29 November–3 December 1982: Abstracts, p. 362, VNIIOFI, Moscow, 1982.

455. V. V. Guseva, L. E. Svyatova, and V. P. Frolov, "Calibration errors in standardizing measurements of the spectral density of irradiance due to nonreproducible positions of the lamps," in: Fourth All-Union Scientific-Technological Conf. on Photometry and Its Applications, Moscow, 29 November-3 December 1982: Abstracts, p. 363, VNIIOFI, Moscow, 1982.

456. G. S. Daletskii, T. P. Dorokhina, M. B. Kagan, M. M. Koltun, V. M. Kuznetsov, I. S. Orshanskii, and V. B. Smirnov, "Optical and metrological parameters of solar cells carried by Venera spacecraft," Geliotekhnika [Applied Solar Energy], no. 2, pp. 7-10, 1983.

457. H. W. Brandhorst, "Variation of solar cell efficiency with air mass," in: Terrestrial Photovoltaic Measurements Workshop Proc., 19-21 March 1975, pp. 51-57, NASA TM-71802, Cleveland, Ohio, 1975.

458. E. S. Rittner and R. A. Arndt, "Comparison of silicon solar cell efficiency for space and terrestrial use," J. Appl. Phys., vol. 47, pp. 2999-3002, 1976.

459. P. R. Gast, "Solar radiation," in: Handbook of Geophysics, Chap. 16.3, pp. 15-16, Macmillan, New York, 1960.

460. N. Robinson, Solar Radiation, Elsevier, New York, 1966.

461. J. F. Allison and R. L. Crabb, "What is AM0? A comparison of CNR and violet cell measurements across the USA and Europe," in: Rec. Twelfth IEEE Photovoltaic. Spec. Conf., Baton Rouge, La., 1976, pp. 554-559, IEEE, New York, 1976.

462. V. Ya. Koval'skii, "Determination of the working parameters of photoconverters working with a solar simulator," Elektrotekh. Prom-st. Ser. Khim. Fiz. Istoch. Toka, no. 7, pp. 19-21, 1971.

463. B. B. Ross, "A new rapid method of solar simulator calibration," in: Rec. Twelfth IEEE Photovolt. Spec. Conf., Baton Rouge, La., 1976, pp. 587-590, IEEE, New York, 1976.

464. V. Ya. Koval'skii, V. E. Kravtsov, L. S. Lovinskii, and L. N. Samoilov, "Simulation of emission spectra using a set of lightguides," in: Third All-Union Scientific-Technological

Conf. on Photometry and Its Metrological Applications, Moscow, 2-6 December 1979: Abstracts, p. 154, VNIIOFI, Moscow, 1979.

465. Yu. M. Belyaev, M. M. Koltun, and V. V. Tsetlin, "Measurement of irradiance on the Cosmos-1129 satellite," Geliotekhnika [Applied Solar Energy], no. 3, 1985.

466. Ya. L. Burtov, Yu. M. Belyaev, Yu. A. Kostenko, and S. P. Tolpenko, "Automatic measurement of the amount of solar radiation," Geliotekhnika [Applied Solar Energy], no. 5, pp. 56-58, 1979.

467. N. S. Lidorenko, S. V. Ryabikov, G. S. Daletskii, V. M. Kuznetsov, M. M. Koltun, and V. B. Smirnov, "Measurement of irradiance in the cloud layer of the planet Venus using solar batteries," Geliotekhnika [Applied Solar Energy], no. 2, pp. 10-12, 1983.

468. V. S. Avduevskii, M. Ya. Marov, Yu. N. Golovin, F. S. Zavelevich, V. Ya. Likhushin, D. A. Melonikov, Ya. I. Merson, B. E. Moshkin, K. A. Razin, L. I. Cherposhchekov, and A. P. Ekonomov, "Preliminary results of an investigation of the radiation regime in the atmosphere and on the surface of Venus," Kosmich. Issled., vol. 14, pp. 735-742, 1976.

469. V. Alekseev and S. Minchin, Venus Reveals Its Secrets, Mashinostroenie, Moscow, 1975.

470. A. D. Kuz'min, The Planet Venus, Nauka, Moscow, 1981.

471. V. I. Moroz, N. A. Parfent'ev, N. F. San'ko, V. S. Zhegulev, L. V. Zasova, and E. A. Ustinov, "Preliminary results of a narrow-band photometric sounding by Venera-9 and Venera-10 in the spectral range 0.80-0.87 μm," Kosmich. Issledov., vol. 14, pp. 743-757, 1976.

472. J. J. Loferski, "Keynote address: photovoltaics 1981 and future prospects," in: Rec. Fifteenth IEEE Photovolt. Spec. Conf., Kissimee, Fla., 1981, pp. 1-3, IEEE, New York, 1981.

473. P. Singh, D. Galley, and E. A. Fagen, "Optical and electrical properties of amorphous silicon-germanium alloy films," in: Rec. Fifteenth IEEE Photovolt. Spec. Conf., Kissimee, Fla., 1981. pp. 912-916, IEEE, New York, 1981.

474. T. Matsushita, M. Okuda, A. Suzuki, et al., "Amorphous Si_xSe_{1-x}—SnO_2 thin film photovoltaic devices: Proc. Sec-

ond Photovolt. Sci. and Eng. Conf. in Japan, 1980," Jpn. J. Appl. Phys., vol. 20, pp. 147-150, 1981.

475. S. F. Pellicori, "Wide band wide angle reflection-reducing coatings for silicon cells," Solar Cells, vol. 3, pp. 57-63, 1981.

476. D. Redfield, "Method for evaluation of antireflection coatings," Solar Cells, vol. 3, pp. 27-33, 1981.

477. D. R. Grimmer and J. G. Avery, "Bonding solar-selective absorber foils to glass receiver tubes for use in evacuated tubular collectors: preliminary studies," Solar Energy, vol. 29, pp. 121-124, 1982.

478. Australian Patent No. 519468. Nickel on Zinc Solar Absorber Surfaces, K. J. Cathro, Appl. 26 January 1978, No. 42828/78, publ. 3 December 1981, MKI F 24 J 3/02, F 28 13/18, C 23 C 3/02.

479. USA Patent No. 4310596. Solar Selective Surfaces, R. van Buskirk, Oral. Appl. 20 March 1980, No. 132153, publ. 12 January 1982.

480. FRG Patent No. 8000783, Vacuum Solarcollector und Verfahren zu seiner Herstellung, H. Limbacher, Appl. 10 January 1980, publ. 16 July 1981.

481. L. P. Randolph, "Photovoltaic outlook from the NASA viewpoint," in: Rec. Fifteenth IEEE Photovolt. Spec. Conf., Kissimee, Fla., 1981, pp. 10-13, IEEE, New York, 1981.

482. R. R. Barthelemy, "Photovoltaic outlook from the Department of Defense viewpoint," in: Rec. Fifteenth IEEE Photovolt. Spec. Conf., Kissimee, Fla., 1981, pp. 14-16, IEEE, New York, 1981.

483. H. T. Jang and S. W. Zehr, "Laser bonding for non-lattice matched stacked cells," in: Rec. Fifteenth IEEE Photovolt. Spec. Conf., Kissimee, Fla., 1981, pp. 1357-1362, IEEE, New York, 1981.

484. E. L. Miller, C. Shy-Shiun Chern, and A. Shumka, "The solar cell laser scanner," in: Rec. Fifteenth IEEE Photovolt. Spec. Conf., Kissimee, Fla., 1981, pp. 1126-1133, IEEE, New York, 1981.

485. J. A. Cape, J. R. Oliver, D. L. Miller, and M. D. Paul, "Automated measurement system for solar cell optical characterization studies of GaAs and multijunction cascade cells," in:

Rec. Fifteenth Photovolt. Spec. Conf., Kissimee, Fla., 1981, pp. 1195-1198, IEEE, New York, 1981.

486. C. G. Hughes, "Silicon photodiode absolute spectral response self-calibration using a filtered tungsten source," Appl. Opt., vol. 21, pp. 2129-2132, 1982.

487. J. S. Hartman, M. A. Lind, and D. A. Chaudiere, "The sensitivity of calculated short-circuit currents to selected irradiance distribution and solar cell spectral responses," Solar Cells, vol. 6, pp. 133-148, 1982.

488. J. R. Hickey, B. M. Alton, F. J. Griffin, et al., "Extraterrestrial solar irradiance variability: two and one-half years of measurements from 'Nimbus-7'," Solar Energy, vol. 29, pp. 125-129, 1982.

489. C. C. Johnson, T. Wydeven, and K. Donohoe, "Plasma-enhanced CVD silicon nitride antireflection coatings for solar cells," Solar Energy, vol. 31, pp. 355-358, 1983.

490. Y. Hamakawa, "Glow-discharge-produced amorphous semiconductors and their application to solar coatings," Thin Solid Films, vol. 108, pp. 304-312, 1983.

491. D. B. Thomas, "An approved laboratory program for photovoltaic reference cell development," Solar Cells, vol. 7, pp. 131-134, 1982.

492. A. L. Fahrenbruch and R. H. Bube, Fundamentals of Solar Energy, Photovoltaic Solar Energy Conversion, Academic Press, New York, 1983.

493. K. L. Chopra and S. R. Das, Thin Film Solar Cells, Plenum Press, New York, 1983.

494. R. Matson, R. Bird, and K. Emery, Terrestrial Solar Spectra, Solar Simulation and Solar-Cell-Efficiency Measurement, Solar Energy Res. Inst., Golden, Col., 1981.

495. M. M. Koltun, "A new standard solar spectrum for measuring the efficiency and electrical properties of photovoltaic installation on the Earth," Geliotekhnika [Applied Solar Energy], no. 1, pp. 18—23, 1987.

496. A. Rohatgi, "A review of high-efficiency silicon solar cells," in: Rec. Eighteenth IEEE Photovoltaic Spec. Conf., Las Vegas, Nev., 1985, IEEE, p. 7—13, New York, 1985.

497. M. A. Green, A. W. Blakers, S. R. Wenham, S. Narayanan, M. R. Willison, M. Taouk, T. Szpitalak, "Improvements in

silicon solar cell efficiency," in: Rec. Eighteenth IEEE Photovoltaic Spec. Cong., Las Vegas, Nev., 1985, IEEE, pp. 39–42, New York, 1985.

498.   R. M. Swanson, "Point contact solar cells: Theory and modeling," in: Rec. Eighteenth IEEE Photovoltaic Spec. Conf., Las Vegas, Nev., 1985, IEEE, pp. 604–610, New York, 1985.